The unburnt egg

more stories of a museum curator

BRIAN GILL

AWA PRESS

First edition published in 2016 by Awa Press, Unit 1, Level 3,
11 Vivian Street, Wellington 6011, New Zealand.

ISBN 978-1-927249-29-1
Ebook formats
Epub 978-1-927249-30-7
Mobi 978-1-927249-31-4

A catalogue record for this book is available from the National Library of New Zealand.

International readers may notice that Māori nouns in this book are not pluralised. As there is no "s" in the Māori language, this is now considered correct linguistic practice.

Cover image: The more damaged of the two moa eggs from Tokerau Beach. Auckland Museum LB4005; photograph by Jason Froggatt.

Inside cover: Plate from "New species of nudibranchiate Mollusca from Auckland waters" by A.W.B. Powell: in *Records of the Auckland Institute and Museum 2(2)*, 1937.

Design by pietabrenton.com
Typesetting by Tina Delceg
This book is typeset in Domaine, klim.co.nz, and Sabon
Printed by Everbest Printing Investment Ltd.

To find out more about Awa Press books and authors, visit awapress.com

Published with the support of

AM

TĀMAKI PAENGA HIRA
AUCKLAND WAR MEMORIAL MUSEUM

THE UNBURNT EGG

ABOUT THE AUTHOR

Brian Gill was curator of birds and other land vertebrates at Auckland War Memorial Museum for over thirty years. He is the author and co-author of many acclaimed books on natural history and has written for *New Zealand Geographic* and *Forest and Bird*. He studied zoology at Massey and Canterbury Universities in New Zealand and held a research fellowship at the University of Queensland in Brisbane, Australia.

ALSO BY BRIAN GILL

The Owl that Fell from the Sky: Stories of a museum curator

Checklist of the Birds of New Zealand, Norfolk and Macquarie Islands, and the Ross Dependency, Antarctica
with Ornithological Society Checklist Committee

The Kiwi and Other Flightless Birds

New Zealand's Unique Birds
with photographs by Geoff Moon

New Zealand Frogs and Reptiles
with Tony Whitaker

New Zealand's Extinct Birds
with paintings by Paul Martinson

Collins Handguide to the Frogs and Reptiles of New Zealand

Contents

From the sea, Auckland—"loneliest, loveliest and last"—
looks like the background of an Italian painting.
A hundred little green conical hills are dotted in the
middle distance against a background of mountains. …
On one of the highest of the little hills above the town
stands a noble building in the Greek style, a modern
Parthenon, looking out across the bay … a thing of
beauty, white against the blue sky.

F.D. OMMANNEY
South Latitude (1938)

Nothing will ever replace the taxonomic knowledge
and training that museums provide; funding in this area
should become a national priority. Otherwise, knowledge
of this planet's biodiversity, and of all the potential
benefits therein, will be lost.

A.V. SUAREZ & N.D. TSUTSUI
BioScience 54 (2004)

Introduction

The stridulations were deafening from the massed cicadas in the trees. I was sweating uncomfortably in my borrowed suit and tie, struggling uphill in the early autumn humidity through the urban bush of the Auckland Domain. I was an outsider—up from south of the Bombay Hills. On the tourist map the museum looked quite close to Queen Street, but the map failed to show the deep gully that intervened. I was lost. A council gardener, removing weeds from the edge of a bush track, pointed me further uphill. Then there came open ground, and I saw it in the distance at the top of the rise: a large magnificent stone building gleaming white in the sunlight and fronted by monumental columns. A palace or temple? A fortress? A prison? I didn't know it then, but to me it was to become all those things for thirty years.

The job interview was in the director's dimly lit workroom. The director, another staff member and two of the museum's governing council sat opposite me across the large work table. They seemed pleased that I had some experience of research on frogs and lizards as well as birds, as the position was to cover curatorship of all the land vertebrates. One of the councillors asked me what I would say to a farmer who rang to complain about magpies. This threw me a bit as I rather like magpies and wish them no harm, but whatever I said must have been satisfactory. A big and a small thing followed: I got the job and—the deception useful when I had needed it most—I almost never wore a suit and tie again.

Three decades in the job amounted to a brilliant vocation. It was a privilege—and a challenge—to be responsible for managing a public collection of 20,000 natural history specimens. Since the collection had built up during more than a century, it included items of intense historical interest as well as scientific importance. I had precious opportunities to conduct research in my fields of interest, especially on the life history of New Zealand cuckoos and songbirds and on the palaeontology of extinct New Zealand birds. I was able to publish my work and was often supported by the museum to attend scientific conferences to present results and hear of the work of others.

Focussing on the needs of the public was satisfying and enjoyable. It included contributing to exhibitions and public events, answering endless public enquiries, and providing access to specimens in the land vertebrates collection for visiting postgraduate students, professional researchers and artists. It was a varied job, and a pleasure to have such a varied group of colleagues—marine biologists, botanists, entomologists, historians, archaeologists and ethnologists, librarians, display artists, security staff, teachers and volunteers, all working under one roof. Then there was the stunning museum building as my place of work and its glorious hilltop location in the midst of a beautiful park.

Biology had thoroughly gripped me at Massey University in the 1970s when I did a degree in zoology. The huge grey science blocks, designed and built by the Ministry of Works, were monumental. There was awe in the number and size of the laboratories stocked with fascinating and futuristic equipment. The dozens of white-coated science lecturers were so knowledgeable in their specialised fields. There was a sense that this was part of the cutting edge where new knowledge was being ferreted out for a greater glory. There was probing and dissection, and the detached, analytical scientific approach. Despite this—and also because of it—the wonder shone through of life's incredible mechanisms and processes, and of living things themselves in all their complex adaptations and riotous diversity.

I came to understand much later that this touches on a point that Charles Darwin made at the close of *On the Origin of Species*, published in 1859. In the book he cast doubt on the prevailing comfortable dogma of a special divine creation, and advanced instead the seemingly grim and desolate idea that lineages of living organisms could change spontaneously and permanently, driven by a brutish "struggle for existence". This was one of the greatest insights of the human mind. Darwin felt that bleak thoughts were not warranted, and he stated that "there is grandeur in this view of life ... whilst this planet has gone cycling on according to the fixed law of gravity, from so simple a beginning endless forms most beautiful and most wonderful have been, and are being evolved."

For much of its history, Auckland Museum's funding levels seem to have been relentlessly poor by today's standards. In the 1930s, for example, newly acquired moa bones were given storage labels written out on the backs of surplus admission cards printed for the museum's public lectures: new blank labels were presumably too great an expense.

When I started in 1982, the large storeroom for land vertebrates was dimly lit by a dozen incandescent bulbs hanging at the ends of cables emerging from the high ceilings. Calico curtains on the large windows were stained and perishing. Many of the mounted birds were stored in rather ghastly cabinets that the museum carpenter had converted from display cases salvaged from the Princes Street museum building, which had closed in 1929. The glass fronts of these recycled cabinets were painted out with pale green or beige. Outside my office was a storage cabinet with a pair of fairly presentable painted wooden doors, but when you opened the doors the interior consisted of a stack of wooden kerosene boxes laid sideways so that each formed an open compartment. Every time you pulled out some of the cotton wool stored there it caught on wooden splinters.

The large extension to the museum's 1929 building, completed in 1960, had been finished as a shell. When the collecting departments occupied these new back-of-house areas, the curators themselves had to paint the walls and ceilings of the offices and storerooms using whatever paint could be bought on sale. This accounted for the colour scheme of battleship grey on some walls and ceilings and a sort of orange-pink on others. In the decade before my appointment stronger finances had been set up, a situation from which I benefited, but it was 1991 before we emptied the entire natural history floor and it was properly fitted out with decent offices and collection stores.

My early years had a certain excitement as I struggled to understand the collection I was responsible for, and ascertain what rightly belonged to it from among potential items in scattered locations. I was concerned to get every last object indelibly numbered and properly registered and catalogued so the specimens and their background information were safely linked.

There was a period when staff passed me boxes of things they thought were "mine", or drew my attention to objects in hidden places. One day the maintenance man showed me a stuffed zebra in a storeroom off the ladies' toilet below ground level near the museum's public entrance. I worked out it had been received in 1907 from the London taxidermy firm of Edward Gerrard & Sons and originally mounted in a group with three kinds of African antelopes. The group had been one of four purchased with funds from the bequest of an Auckland solicitor and local politician, Edmund Mackechnie, who died in 1901. When I first saw the zebra one of its ears was missing; a replacement ear was fabricated in 2002 when we had a taxidermist on the staff.

Another project was climbing a ladder to a very dirty area above a storeroom for ethnographic carvings and retrieving dozens of mounted trophy heads of deer and other animals. I had to wipe away dust and search each head carefully for labels or inscribed numbers that would be the key to unlocking their background.

Another time I was called to a hidden and disused staircase—a space currently occupied by a kauri tree exhibit in the Land Gallery. The staircase had once served the back of the 1929 building and had lovely ornamental metal banisters with wooden handrails. Up and down the stairs that connected three floors there were abandoned display props and pieces of broken furniture. My task was to examine a big board on which was mounted a large python skeleton with broad sinuous curves. There was no skull and half the other bones were missing or now lay in an untidy heap on the floor below. I salvaged a selection of ribs and vertebrae, cleaned them, and reunited them with a large python skull and mandibles in the main collection that seemed to be from the same animal. I later realised that this snake skeleton, received on exchange in 1878 from the Muséum national d'Histoire naturelle, Paris, is one of the oldest surviving specimens in Auckland Museum's reptile collection.

In the land vertebrates department I was surrounded every day by animal specimens and the implements and equipment needed to prepare and study them. I became habituated to things that to most outsiders seemed odd and quirky. One departmental curiosity was Gilbert Archey's osteometric board. Archey had been director of the museum from 1924 to 1964. His paper *The Moa—A Study of the Dinornithiformes*, published in 1941, was the museum's first scholarly "bulletin" and it had won him a doctorate of science from the University of New Zealand.

The publication has a diagram of the device Archey used to measure the dimensions of moa leg bones, and because the instrument in the department is similar, and old, I assume it was his. It comprises a plank of wood more than a metre long. Moa bones are large—the tibiotarsus (the bone in the chicken's drumstick) of the tallest moa was probably the longest bird bone ever. A wooden ruler—marked, fortunately, in centimetres as well as inches—is recessed flush along the length of the device. Lined up exactly at the zero end of the ruler is a wooden stop, resembling a bookend, against which you can place one end of the bone to be measured. An opposing "bookend" is mounted on a strip of wood that runs freely the length of the device in a slot. You bring this moving stop against the other end of the bone and the bone's length can be read off where the moving stop crosses the ruler. It is a perfect instrument for measuring long items to the nearest millimetre or two and I used it regularly.

Another prominent item was LM131, an articulated human skeleton mounted upright in a standing position. Too tall for a cupboard, the skeleton stood on the sidelines in the collection storeroom. The museum had received it in 1883 by exchange from Henry Ward of Rochester, New York, who ran a biological supply company; the company still operates as a supplier of biological specimens to institutions worldwide for teaching and display. I often used the skeleton to confirm or rule out that a particular unidentified bone was human. Having the bones assembled in their correct positions is a great help for naming the bones and determining (for paired bones) whether they are from the left or right side. Being able to touch real bones is important because the look and feel of the surface texture helps to identify fragments.

Besides its use as a reference specimen, LM131 is historically important as an example of Victorian techniques of articulation. Whereas the small bones on more modern skeletons are often glued together, the bones on LM131 are finely drilled, where necessary, and held together by wires and metal pins and bolts. It's a beautiful object. The octagonal wooden base is painted black, and from its centre arises a strong metal rod (with an ornamented attachment ring) that passes between the legs and through the pelvis to support the spine.

Around 1898, Thomas Cheeseman created the "Blue Book", the museum's first register of vertebrate specimens. It's a large bound volume, "folio" in size at around 380 by 245 millimetres, with column headings for details of each specimen. The first specimen Cheeseman entered, the particulars in dark blue ink, was the human skeleton. Its number was therefore V1 ("V" for vertebrates) and this is still visible in original lettering on the base. (LM131 is its number in the current numbering system for the museum's mammal collection.)

The skeleton is short in stature, but it belonged to an adult because it has fully erupted wisdom teeth. Features of the pelvis suggest it is a male. In a letter to Ward dated December 4, 1882, Cheeseman had requested a "well-mounted human skeleton—adult and perfect, and if possible, a male". He wanted it mounted in the way that was most suitable for museum purposes, and added, "I should prefer one of the North American Indians." Ward had the skeleton despatched early in 1883; it was exchanged for kiwi eggs and skeletons.

Victorian museums collected human bones, especially skulls, because there was scientific interest in the physical variability of humans. Big museums assembled large collections from around the world, including from all regions of Europe. Much contemporary research on variations of the human skull proved unreliable (and some was misguided), but the collections have supported later, sounder research, such as recent computer-assisted investigations of relationships between skull shape and geographic origin.

People seldom got exactly what they asked for in the late nineteenth-century museum exchanges, and Cheeseman didn't get a native American skeleton. In the Blue Book column headed "Locality", which Cheeseman always filled out, he wrote against the entry for V1: "Locality not known." It was presumably a regular skeleton of the kind prepared for medical and teaching purposes. The museum's zoology staff have cared for LM131 for 130 years, treating it with respect. With its original mounting, long institutional history, and association with the famous American supplier, LM131 is an important artefact in the museum's rich history.

In 1986 I asked at a curators' meeting whether we ought to have a new staff photograph. The director agreed and asked me to organise it. I turned to the Yellow Pages to find a suitable photographer, settled the details and put a memo in staff pigeonholes advising the date and time. On the day it was sunny, and everyone assembled on the museum's front steps. The photographer had checked the site and suggested we move down to the steps in front of the nearby cenotaph so the building would be a backdrop. When I received the proof photos I found they had come complete with a caption. Just above the heads of the back row of smiling staff members was the large inscription "The Glorious Dead" engraved upon the cenotaph. That was one staff photograph that never made it to the pages of the *Annual Report*.

Building natural history collections is a slow process. Small, steady additions make for a series of specimens that are distributed historically as well as spread geographically. After more than a century of natural history collecting by New Zealand museums, researchers are in a good position to study changes in organisms over time.

An example is the study-skins of the sacred kingfisher, *Todiramphus sanctus*. This is a common bird in New Zealand, and Auckland Museum now has seventy-six study-skins, most from the northern half of the North Island. The museums in Wellington, Christchurch and Dunedin also have specimens, and these are mostly from their own regions, so the collections complement each other. To study kingfishers across the whole country a researcher needs to visit all four collections, and it is a safeguard against calamity that the nation's bird collections are distributed in this way. The same species of kingfisher occurs in Australia and nearby regions, and Auckland Museum's collection has five foreign skins, a useful reference point for New Zealanders considering geographic variations in this bird.

Four of the local kingfishers were collected in Auckland suburbs in 1878, 1884 and 1897. Not all were shot—on one label the collector is recorded as "The Cat". There are no specimens from the 1920s, but every other decade of the twentieth century is represented by between three and fourteen birds. Since the 1950s, Auckland Museum's bird specimens have been prepared from birds found dead and handed in by members of the public. Flying into window glass, striking cars or being killed by domestic cats have been the main causes of death. Five kingfisher study-skins were added in the first decade of this century.

By preserving such natural history specimens, museums often save researchers the time and expense of collecting their own samples in the field. The aggregation of specimens into public natural history museums is a convenience for researchers that often translates into savings in taxpayer funding. Access to museum specimens is enhanced by electronic versions of museum collection catalogues, increasingly available to be searched online from personal devices anywhere in the world. The maintenance and storage of the collections is inexpensive, given the big picture of the long-term benefits and cost-savings.

A recent example of the benefit of museum specimens to researchers and to taxpayers involved two biologists from Otago University, Lisa Tracy and Ian Jamieson, who in 2011 published in the journal *Conservation Genetics* a paper about an endangered New Zealand bird called the yellowhead. The few remaining scattered yellowhead populations show some distinct genetic differences. Tracy and Jamieson compared DNA from contemporary yellowhead blood samples with DNA extracted from tiny slivers of tissue from the toe pads of museum specimens. At three New Zealand museums, and seven museums in Australia, Britain, Canada, the United States and Austria, there were a total of sixty old yellowhead specimens, many from places where the species is no longer found. Auckland Museum contributed eight of the specimens.

The museum DNA established that the original, widely distributed populations of yellowheads lacked significant regional genetic differences. The genetic variation seen today must be a random effect of the severely reduced populations. This is a pleasing result because it means that in moving yellowheads to predator-free locations to try to save them from extinction, the Department of Conservation does not need to save all the surviving sub-populations, nor keep them separate in an attempt to preserve their genetic diversity. The birds from any source can be mixed, making the task easier and cheaper and saving taxpayer money.

This is all by way of introduction. What follows is a collection of stories about various natural history specimens that I came to know, and the history, people and science that lie behind them.

A hapless king penguin

On June 10, 1885, Charles Francis Adams, a young American taxidermist who had moved to Auckland five months earlier to work at the museum, was woken about midnight by noises in his boarding house. "I looked out of my window and saw the air full of sparks of fire," he wrote next day.

He dressed quickly and stepped into the hallway. People were hurrying out with their furniture. This "served as a hint for me to move my effects". He helped move Mrs Murray's furniture. Poor Miss Sampson "was so terrified that she is partially paralyzed as a consequence". The fire was put out before it reached Adams' part of the wooden building but he wanted to sleep elsewhere. Knocking at the door of another boarding house but getting no answer, he walked to the museum. There, presumably in his basement taxidermy workshop, he "built a fire and lay on [his] blankets on the floor beside it".

These charming details are in a daily diary that Adams kept of the first six months of his employment at Auckland Museum. The diary gives an insight into the working life of a Victorian taxidermist in New Zealand and helps to document the early years of the museum. Founded in 1852, it occupied its first purpose-built building in 1876. This was in Princes Street, close to the centre of town, on a site now occupied by a high-rise hotel. It comprised a main exhibition hall at street level with a mezzanine gallery above. Besides Adams and the curator, Thomas Cheeseman, the only other employees were an attendant employed on Sundays and a janitor. Adams worked from Monday to Friday and on Saturday mornings; his salary was £200 a year.

I discovered his diary—or rather a brief catalogue entry for it—in the library of the University of Illinois at Urbana quite by accident in 2011, while I was trying to understand the workings of internet searches by half-heartedly entering names of interest. I found the library's website and looked for the email address of a librarian. I was greatly helped by John Hoffmann, who answered my request.

In 1974 the descendants of Charles Adams' family had offered the diary for photocopying by the university, which was in the nearest city to the farm where Adams had grown up. John Hoffmann posted me a copy of the original photocopy so I could read the diary in full. The university was pleased to have the diary examined by someone who could interpret its content, and I provided a description of it for their cataloguing system. Hoffman also sent me a copy of Adams' alumnus record: Adams had studied taxidermy at this university, as well as at Henry Ward's biological supply house in Rochester, New York.

After I started at Auckland Museum C.F. Adams was one of the hundreds of names of collectors or presenters of specimens that I came to know as I became ever more familiar with the land vertebrates collection and its documentation. He was listed as the source of several American bird specimens and mammals from Borneo. I knew many names of donors, but in most cases had no idea who the people were.

In the late 1990s, in Auckland Museum's archives, I came across five letters from Adams to Cheeseman, and found Cheeseman's replies in the museum's letter-books—the duplicated copies of outgoing letters. I could see from the context that Adams must have been the unnamed American taxidermist whose employment was alluded to in the museum's annual reports at that time. Adams' letters had been written after he left Auckland, and he reported his latest activities to Cheeseman in a spirit of respectful friendship. Cheeseman's letters were headed "Dear Mr Adams" and the replies were signed "C.F. Adams", so at this stage even Adams' first name remained a mystery to me.

Around 2005 I was invited to write a chapter on Auckland Museum for an American book on the history of Australasian bird collections. The editor saw my sub-heading "C.F. Adams" when the names of other taxidermists were given in full, and on his own initiative searched for and found an obituary of Adams in an American bird journal. At last I had the full name: Charles Francis Adams. Then, unexpectedly, there came the diary, with its detail on the minutiae of Adams' life in Auckland and elsewhere.

After the delight of the diary I received further news from John Hoffmann. He had tracked down a photograph of Adams that had been taken in connection with the university's alumni records. An email with an attachment arrived. This was a big moment. I paused, double-clicked, and there he was. From C.F. Adams of the museum registers, to Charles Francis Adams of the obituary, to the neat handwriting of the diary, I could now see his likeness. The photograph was a studio portrait of a handsome young man with neat short hair and a moustache long enough to obscure his upper lip. He seemed every bit the Adams of the diary—sober, thoughtful and conscientious, diligent in his work and aiming high in his standards.

Adams left his home in Champaign, Illinois for New Zealand on December 10, 1884 at the age of twenty-seven. His train journey to San Francisco took eight days because he travelled on the narrow-gauge Denver and Rio Grande Railway through mountainous Colorado, rather than on the more direct Union Pacific line to Ogden in Utah. He may have been taking the opportunity to see new country. His diary is full of notes on scenery, landforms, vegetation, railway engineering, the evidence of extractive industries such as coal shafts and stone quarries, and cattle ranches. A desire for even more travelling was probably why he would leave Auckland Museum after just two years, collecting birds and mammals in Borneo. His letters to Cheeseman also reported on a trip he made to the Galápagos Islands, and he had ambitions for further travel despite suffering from seasickness.

From San Francisco, Adams took the mail steamer *Zealandia* across the Pacific to Auckland, departing on Christmas Eve. The ship, a single-screw steamer with one funnel, was also rigged for sail with four masts and had a speed of thirteen knots. Built in Glasgow in 1875, it carried up to 300 passengers (in three classes) and had British officers and a Chinese crew.

The voyage started badly. "I was taken miserably sick and went to the side to vomite [sic]. The rain wet me from above and the waves broke over the deck so my feet were soaked with water." On Christmas Day he felt dizzy and could eat nothing and the misery continued for two more days. On the 28th he felt better but the drinking water was poor and he was very thirsty. Seasickness returned next day and he was pessimistic about the rest of the trip, wondering if he was wise to have travelled in steerage. His spirits lifted on the 31st: he could still not eat a "hearty meal" but the weather was fine.

When he arrived Adams was favourably impressed by Auckland, noting: "The city looks quite nice as the buildings are new looking which gives a look of freshness." He was impressed with the cleanliness of the streets, which were sprinkled from a nozzle screwed on to a hydrant: the water moved the dirt along. On his first Sunday he went walking and was "much pleased to see fine flower beds in the front of residences": "The city park [Albert Park] is very nicely kept and being situated higher than a part of the city a splendid view is to be had from it."

There were worries though. One evening a fire broke out in the marble works near a rope-making factory. "It seems that a great many fires occur for the size of the place as the alarm is sounded almost every night. The hills keep the firemen [presumably with hand-hauled equipment] from arriving quickly on the spot."

Adams noted that he was required by the museum to assist in taking "readings" daily, including on Sundays. He often referred to going "up" to read the instruments. Auckland Library has a magnificent photograph of Princes Street taken in 1902 by Henry Winkelmann, which is reproduced opposite page one of this book. It shows the museum building side-on from a higher vantage point. On the roof, behind the façade, wooden steps lead up to a small platform on which are visible a Stevenson screen—a louvred enclosure to shelter meteorological instruments from rain and direct sun—and an anemometer, consisting of rotating arms with hemispherical cups at the ends, for measuring wind speed.

The letter-books record that about every month Cheeseman would send a note to James Hector, director of the Colonial Museum in Wellington, listing meteorological data. Adams' diary helped to recover the fact, long forgotten, that Auckland Museum had run a weather station in the 1880s, probably as part of a national network.

Tallying up the specimens mentioned in the diary shows Adams prepared at least 124 birds, fish, reptiles and marine invertebrates in the first six months. Many of the birds were foreign and with these he must have been finishing off specimens the museum had received as roughed-out skins from overseas suppliers. Notable New Zealand species that Adams processed included a New Zealand "thrush" or piopio (probably *Turnagra capensis*), now extinct, which he mounted in January 1885, and "owl parrots", or kākāpō (*Strigops habroptilus*) that he worked on later in the year. In April 1885 someone brought in a blue-wattled crow, or kōkako (*Callaeas wilsoni*), that had been killed in the Waitakere Ranges near Auckland.

Adams spoke to Cheeseman about using the workshop after hours to build up a personal collection of birds. In due course he could take these back to America for sale. Cheeseman had been "very obliging indeed", his main concern being that Adams "be careful about turning the gas out".

In 2011 I visited the Smithsonian Institution in Washington, D.C. to look at birds that Auckland Museum had sent there on exchange in the 1880s. Among these were several study-skins with plain manila-coloured labels in Charles Adams' handwriting. He would have prepared these during his paid hours at Auckland Museum and they would have been selected by Cheeseman as currency in the exchanges.

Just from the label I could tell if a specimen was from Auckland Museum. At the time the museum used manila labels (they seemed to be commercial luggage labels) that had a dull pink eyelet to strengthen the hole for stringing, and some characteristic printed lettering from the manufacturer. Another clue was that the order of the information on Adams' labels was always Latin name, sex, common name, location, date (with a slash between the month and year) and eye-colour, as in the following example from a label attached to a tūī—a kind of honeyeater—in the Smithsonian collection:

Prosthemadera novae-zealandiae
♀
Tui
Motu Tapu [Motutapu Island near Auckland]
Aug. / 85
Iris brown

Also among the Smithsonian tūī I found a study-skin with a preprinted Adams' label inscribed "C. F. ADAMS' NEW ZEALAND COLLECTION". This had to be a bird from Adams' private collection, which he had sold directly to the Smithsonian or a third party. Many New Zealand birds that Adams prepared while at Auckland Museum remain in the collection there, but others will have been distributed to museums around the world by exchange from Auckland Museum or through the disposal of Adams' own collection.

On Tuesday, May 12, 1885, while Adams and Cheeseman were upstairs in the main part of the museum, a man stole a copy of Walter Buller's *A History of the Birds of New Zealand* from the taxidermy workshop. Adams would have used the coloured plates in this book as reference for his taxidermy work. Two days later a detective reported the stolen book had been recovered. Next day another detective advised that the thief had been apprehended.

Adams and Cheeseman had to attend court several times,

but at his appearance on July 6 the thief pleaded guilty and neither man was called as a witness. The *Auckland Star* reported how the story unfolded:

> A LITERARY LARCENY. – George Edward Dalton, alias Henry J. Atkins, was charged with the larceny of Dr. Buller's work on New Zealand birds, from the Auckland Museum, on the 12th of May, 1885. – Prisoner pleaded guilty, and stated that he had taken the book, not being aware of its great value. He was destitute at the time, and acted on the impulse of the moment. – His Honor [sic] stated that as this was not the prisoner's first offence, he must sentence him to six months imprisonment with hard labour.

Living in New Zealand, Adams could not avoid hearing about Britain's imperial wars. He noted on February 13, 1885 that a topic among his circle was "the condition in Soudan": General Gordon had been shot dead by rebels at Khartoum on January 26. The following month the focus had shifted to the "Russian Scare"—"People here are very much excited over the trouble with Russia and defensive measures are being taken. ... About noon [on April 10] a telegram announcing the crossing of the front, by the Russians, was received [by the newspapers] and caused considerable excitement."

On April 22, Adams noted Auckland's increasing preoccupation with the prospect of war, and five days later he called on the US Consul in Auckland to get reassurance he would not be liable for any New Zealand military duty. On the way back from collecting marine invertebrates on the North Shore, Cheeseman and Adams stopped to see the "defensive works" on North Head: artillery batteries were being built to defend the port.

In his spare time Adams often walked to the wharves, a great attraction in a port city, and saw the international shipping that was the colony's lifeline to the outside world. He marvelled at the technology. Aboard the steamer *Doris*, which had arrived from Britain with immigrants, he was impressed it was "lighted with electricity". On the *Waimate* he looked through the galley and was shown the apparatus for making freshwater from sea water. There was a poignant moment one Sunday in June 1885 when he again saw his "own" ship *Zealandia*. As the vessel departed into the darkness he "turned away and thought of her long trips and how many prayers have been offered up for her safety, as she ploughs the seas, by dear friends left behind".

The diary entries stop abruptly at July 12 that year, as if Adams were suddenly tired of making a detailed record. After this there are three isolated accounts of local trips from Auckland to offshore islands. "Trip to Kerewa [Karewa] Island" covers a visit in December 1886 to this tiny island that lies about six kilometres off Tauranga in the Bay of Plenty. The island was well known for its tuatara, lizard-like reptiles that lived in association with burrowing seabirds. Adams' objective was probably to collect the specimens he later used to construct a small diorama at the museum, where the tuatara were depicted at a burrow.

Reaching the island aboard SS *Clansman*, he was put ashore on one of the ship's boats. "The Capt. kindly helped me to fit out properly for my stay of one week alone." The island, he reported, "was little more than a large rock (2 or 3 acres) with one peak and the greater part was covered with a low scrubby bush. On going ashore my attention was first attracted by the great number of green leaves on the ground. Underneath the bushes the ground was quite bare of grass and full of burrows of a species of Puffin." These were probably flesh-footed shearwaters, *Puffinus carneipes*, which still breed today on the island.

"I first cleared away some vines and pitched my tent, which had to be done over some of the birds' burrows. As the sun neared the horizon Puffins approached from seawards and circled widely about the island, some collecting in rafts on the surface of the water. Just at dusk they came closer in and commenced diving into the bushes, which at once explained how the leaves had been knocked off."

After dark, Adams took a lantern and a bag in search of tuatara, which stood still in the lamplight and were easily seized by the back of the neck. The commotion of the shearwaters calling and crashing about in the vegetation was, he wrote, a "strong contrast with the quiet condition during the day when a casual observer would think there were no birds to be found there at all".

The week alone passed pleasantly enough. The weather was fine and Adams found the birds and tuatara "sufficient company to keep me from getting lonesome". The shearwater eggs he found in the birds' burrows proved "very fine eating".

On March 20, 1885, Cheeseman had received a telegram from a Dunedin taxidermist, William Smyth, offering a live king penguin from Macquarie Island, south-west of New Zealand. This arrived ten days later on SS *Wairarapa*. Smyth had put the penguin in a padlocked box and arranged for the ship's butcher to feed it en route for a consideration of five shillings.

Cheeseman was instructed to feed the penguin small fish and raw lean meat. "It has no bad nature," Smyth observed. "If it snaps at one's finger it means nothing; it will not injure anyone's finger and will soon know and follow whoever feeds it."

King penguins, *Aptenodytes patagonicus*, stand almost a metre tall and are second in size among living penguins only to the closely related emperor penguin. They are roughly circumpolar in distribution, breeding north of the pack ice on Antarctic and subantarctic islands. Macquarie Island has more than 70,000 nesting pairs.

With the penguin at the museum, Adams noted: "It is very amusing and took some fish which I bought for it tonight." An art school was run at the museum, making use of exhibits such as the plaster copies of classical statues, so naturally "the young ladies of the drawing class came down in a body to see the penguin". In Auckland there was no zoo where people could see live animals, and the many visitors who saw the penguin at the museum were "very much amused by its movements. … It devours fish greedily and always seems to be hungry."

Adams took the penguin to the studio of Josiah Martin, one of Auckland's leading photographers, so images could be made by the "instantaneous method"—a way of capturing movement using flashes of short duration. Then on April 9 there was a peremptory diary entry: "This evening I chloroformed the penguin." Next day he started to prepare from it both a skin and a skeleton.

Many of the specimens that Charles Adams prepared during his two years at Auckland Museum were immediately put on display to educate and delight the public. A few remain on display 120 years later. An articulated skeleton of a king penguin is probably Adams' unfortunate bird: registration details show it was received in 1885 from Smyth and had come from Macquarie Island.

The study-skins prepared by Charles Adams have been noted and measured over the decades by a host of visiting scientists and students collecting data for research projects. This is the legacy of the quiet diligent man from Illinois who crossed half the globe to provide much needed taxidermy services to a small but growing colonial museum.

Secrets of the shining cuckoo

On a hot sunny day in November 1978 I was lucky in my field observations. As I walked through the dry kānuka forest at Kōwhai Bush near Kaikoura (in the north-east corner of the South Island of New Zealand), the air seemed to hum with the thrill of summer. Above the din of the cicadas I heard the distinctive whistling call of a shining cuckoo and went in pursuit.

Parasitic cuckoos have simple calls, most likely because their vocalisations need to be largely innate: while many young birds learn refinements of their calls from their biological parents during their period of dependency, this cannot be so in those cuckoos where the female lays her eggs in the nests of another species and plays no further part in raising the young. However, although the shining cuckoo's loud, clear call may be simple, something about it makes the position of the calling bird hard for the human ear to locate. The first iterations of the repetitive call may seem quite far away, but then the call seems to get louder and closer and the bird may even turn out to be perched nearby.

The shining cuckoo weighs around twenty-three grams—about three-quarters the weight of a house sparrow—and the sexes are alike in plumage. The bird belongs to a group known as the glossy cuckoos for their iridescent plumage. The dozen or so species in this group are the smallest members of the cuckoo family and inhabit wide areas of sub-Saharan Africa, Asia, New Guinea and Australasia.

Shining cuckoos are very hard to see in trees or scrub, despite—or perhaps because of—their striking appearance. The plumage of the upper parts is dark lustrous green, shot with copper or purple according to the angle of light. The under parts have bold dark green stripes running from side to side on a white background. The bird that seems so gaudy in the hand is remarkably cryptic perched still on a branch. Eileen Duggan captured this in her 1929 poem *The Pipiwharauroa* (*Shining Cuckoo*):

> And I burnt my eyes with gazing.
> Still I see the poplars shiver,
> Still I hear the little runnels down the folded gully falling,
> But I never saw the bird!

I persevered that day at Kōwhai Bush and eventually found two shining cuckoos. It is reward enough to watch such transient and secretive birds, but to cap it off I saw a red colour band on one of the birds' legs. Parasitic cuckoos tend to have short legs that are tucked well in when they perch. The feathering extends much further along the leg than in songbirds, so it is difficult to see the legs clearly to check for bands.

I lost sight of the cuckoos that day, but ten days later I found them in the same patch of bush. After half an hour of careful stalking I approached to five metres away—closer than my binoculars would focus—and there was no doubt that one cuckoo had a metal band on its left leg and a red plastic band on its right. This was B-40201, a cuckoo I had banded as a nestling exactly two years before at a grey warbler's nest a kilometre away. In my three summers of studying shining cuckoos at Kōwhai Bush I had been able to band only eleven nestlings from the small number of parasitised warbler nests I had under observation. I had banded no adults, but not for lack of trying to capture them by mist netting. I had given each nestling a unique combination of one numbered metal ring and up to two plastic colour bands. A numbered ring serves as an indelible marker should a bird be re-caught alive or found dead; the colour bands allow individual birds to be recognised at a distance from the exact combination on the bird's legs.

This young cuckoo had no doubt departed New Zealand at least once on its autumn migration north to the tropics, and had now returned to its hatching place attempting to breed. In the last sixty years only around 120 shining cuckoos have been banded in New Zealand, so this recovery record—with the suggestion of homing to a natal area—was important data.

Almost exactly 200 years before my fortunate field observation of a marked cuckoo at Kaikoura, and in the same month, others had been excited by the bird. The first shining cuckoo known to science was shot in November 1773 from aboard HMS *Resolution* at Queen Charlotte Sound, just 150 kilometres north of Kaikoura, during Captain James Cook's second voyage of discovery. From among the day's haul, perhaps the father and son painters Johann and Georg Forster found the tiny bird a particularly exotic and amazing subject, given its shining colours and bold pattern.

Back in Britain, the author and naturalist John Latham used the Forsters' painting and accompanying notes to list the species as "shining cuckoo" in his *A General Synopsis of Birds*, the part in question published in 1782. However, it was not properly named, consistent with today's rules of zoological nomenclature, as *Cuculus lucidus* until 1788, in a publication by a German naturalist, Johann Gmelin. Hence the bird's full zoological name today is *Chrysococcyx lucidus* (Gmelin, 1788). The brackets around the naming authority signify that it is now in a different genus from that in which it was originally placed. Things in zoological nomenclature often change and, to make the shining cuckoo's naming even more complicated, the Asian and Australo-Papuan glossy cuckoos are sometimes put in *Chalcites*, leaving *Chrysococcyx* for the African representatives.

Three of the shining cuckoo's habits are especially remarkable. One is its brood-parasitic breeding. Parasitic cuckoos lay their small eggs individually in the nests of foster parents smaller than themselves. The cuckoo egg hatches quickly and the cuckoo chick grows rapidly, pushing any eggs or other nestlings out of the nest. Being larger than the foster parents' own chicks it needs to be reared alone so it can monopolise all the food they supply.

Another notable feature is the shining cuckoo's diet. Among the insects it eats are many that other birds find distasteful. I studied the stomach contents of a sample of twenty shining cuckoos from the South Island that had died accidentally. Half the stomachs contained caterpillars—up to sixty per stomach—and a third of these belonged to the magpie moth, whose caterpillars are covered in sharp defensive hairs. Clusters of the caterpillars' spines in regular rows had pierced the soft gizzard lining, in much the same way the bristles of a brush are inserted. It's possible the cuckoo sloughs off the gizzard lining, complete with caterpillar hairs, and coughs it up. More than half the cuckoo stomachs held beetles—up to fifty-five per stomach. Nearly ninety percent of them were ladybirds. These insects discharge a poisonous fluid when molested and their colour is supposed to warn off birds from eating them.

A third wonder is the shining cuckoo's migration. Over the years I helped to work out the details of this by studying specimens in natural history museums. Shining cuckoos breed on mainland New Zealand and the Chatham Islands, where they parasitise the grey warbler and Chatham Island warbler respectively. The same species of cuckoo also breeds in south-west and south-east Australia. There it is called the shining bronze cuckoo and parasitises various birds that are close relatives of New Zealand warblers and, like them, build covered pear-shaped nests suspended in the foliage of trees and bushes. Shining cuckoos also breed in New Caledonia, Vanuatu and Rennell and Bellona Islands at the southern end of the Solomon Islands chain, and in all these places the host is—or is assumed to be—a local warbler with a covered nest. The populations breeding on these tropical islands are non-migratory.

Shining cuckoos that breed in the temperate latitudes of Australia and New Zealand are absent there in winter. It's thought they migrate north to spend the season in the tropics, the band of territory that runs from the islands of eastern Indonesia through New Guinea to the Solomon Islands. A study by the great German-American bird curator Ernst Mayr, published in 1932 and using specimens newly collected by the Whitney South Sea Expedition, showed that birds belonging to the New Zealand subspecies—with wide bills on average—seem to go east of New Guinea to the Solomon Islands. Since an article by New Zealand biologist Barry Fell in 1947, it is commonly assumed that shining cuckoos migrate directly between New Zealand and their wintering grounds, with the only possible landfalls en route at Norfolk and Lord Howe Islands. This constitutes a transoceanic return journey of 6,000 kilometres.

The Whitney South Sea Expedition was a venture of the American Museum of Natural History in New York and sponsored by wealthy philanthropists Harry Payne Whitney and Gertrude Vanderbilt Whitney. From around 1920 to 1932 a sailing boat systematically travelled to a myriad of islands across the south-west Pacific, collecting birds at each location. Subsequent study and comparison of the specimens back at the museum graphically showed that many types of birds had spread widely in the region, and on separate islands they had formed populations that differed in outward appearance. The island populations, although they clearly had a common ancestor, were now isolated from each other and were diverging in form and structure and evolving into separate species. This wealth of new data helped to refine the twentieth-century reinterpretation of the mechanics of evolution.

When the Whitney expedition visited Western Samoa in 1924, New Zealand was the colonial power. A condition of the collecting permit was that a representative selection of prepared and labelled study-skins of Samoan birds be returned to New Zealand authorities. Several of these important and historic Whitney specimens—mostly collected and prepared by Rollo Beck, a notable American taxidermist—are held in the bird collection at Auckland Museum. Natural history collections may grow and diversify in random opportunistic ways like this, as well as through systematic, focussed collecting.

But I am straying from the shining cuckoo's story. The Australian subspecies, which is characterised by birds with narrower bills on average than the New Zealand birds, seems to winter in New Guinea and the Lesser Sunda Islands at the eastern end of the Indonesian archipelago, and crosses continental Australia to get there. Although the technology is improving, shining cuckoos are still too small to carry satellite-tracking transmitters. None of the shining cuckoos banded in New Zealand have been recovered overseas, so we lack precise details of the migratory journeys. Instead we rely on bill widths to deduce that Australian and New Zealand birds have separate tropical distributions, paralleling their separate east-west breeding grounds.

To retest this assumption about separate migrations I undertook to examine study-skins of shining cuckoos held at natural history museums around the world. In any scientific study you need to set your protocols. For New Zealand and eastern Australia I confined my attention to the fresher and more reliable specimens that had been collected after 1944. For Western Australia and the tropics, samples were small so I used specimens of any vintage, the oldest of which was collected in 1868. I used only adult specimens so the bills would be fully developed, and rejected any birds with damaged bills.

In Brisbane, where I was employed at the start of the project, I examined and measured the shining cuckoos in the Queensland Museum, and the curator of birds kindly arranged to borrow specimens from museums in Perth, Hobart, Launceston and Port Moresby so I could see them too. Around this time I travelled to Melbourne, Sydney and Canberra for other reasons and was able to visit the museums there to measure their shining cuckoos. A holiday trip to London and Paris had me studying more cuckoos, and after later returning to New Zealand to live I gradually ticked off the birds held at the museums in Auckland, Wellington, Christchurch and Dunedin. Meanwhile, two helpful curators at museums in New York and Leiden sent me measurements they had made of the shining cuckoos in their own collections.

My main interest was in the width of each bird's bill, measured to the nearest tenth of a millimetre just in front of the nostrils. This varied from 4.1 to 6.1 millimetres. I compiled a set of data based on some 480 cuckoo specimens, with each measurement linked to the bird's place and month of capture. My graphs of the frequency with which specimens had been collected in different months and in different regions reconfirmed earlier assumptions that the Australian and New Zealand shining cuckoos inhabited the temperate and tropical zones at complementary times.

The bill-width measurements supported previous notions of the migratory destinations. The average was 5.2 millimetres in the Bismarck Archipelago (New Britain, New Ireland, et cetera), 5.3 in the Solomons and 5.4 in New Zealand, suggesting the same population was involved in migrations between these destinations. Likewise, mean bill widths of 4.5 millimetres in the Lesser Sundas, 4.4 in New Guinea, 4.5 in Western Australia and 4.8 in Tasmania suggested a separate north–south interchange of birds between those areas.

But I found that something strange was happening in eastern Australia. All cuckoos from Tasmania and Victoria were narrow-billed, as expected, but I discovered a large proportion of wide-billed birds among the narrow-billed ones collected in New South Wales and Queensland. When I looked at seasonality, by plotting bill width against month of collection, the penny dropped: wide-billed birds were present in New South Wales and Queensland only during the months of migration.

The data seem to confirm the proposition, first suggested by Ernst Mayr in 1932, that some, perhaps all, of the New Zealand cuckoos migrate via eastern Australia. This extends the return journey of New Zealand birds to Melanesia to about 12,000 kilometres but involves shorter distances across water—just over 2,000 kilometres between New Zealand and Australia, and just under 2,000 between Australia and the Solomons region. Of course it also allows the tiny cuckoos to rest and feed as they move up and down the Australian coastal plain.

It was sobering to think of the effort that had gone into the sample of 480 shining cuckoo study-skins to which I had access. Some of these birds would have been shot during expeditions to what were then remote locations and under trying conditions. Others would have crashed against windows, or been brought in by the cat, and the bodies carefully retrieved and taken to local museums by thoughtful citizens. Each bird would then have had to be skinned and stuffed by a preparator, a complex and exacting task. Each labelled specimen then had to be stored in a museum, and cared for during handling by researchers, building renovations, relocations of collections and even (in some cases) war. But in the end, the careful work of perhaps a hundred collectors, taxidermists and curators operating at a hundred locations during a hundred years had given me a representative sample of shining cuckoo study-skins. From these I was able to tease out a few of the bird's secrets, all to satisfy human curiosity and our burning desire to understand the natural world.

The shining cuckoo's spectacular migration to and from New Zealand is remarkable for the smallness of the bird and the vastness of its over-water journeys. But much is still unknown. As Eileen Duggan wrote:

> Are the seas to you as homely as our fields of curling
>   clover?
> What old memory sends you blindly over hill and
>   over hollow?
> Do you never doubt the way?

The unburnt egg

One night in February 1961, in the city of New Plymouth on the west coast of the North Island of New Zealand, an old wooden house caught fire and its occupant was killed—a woman of eighty named Blanche Halcombe. This tragedy had a connection to the land vertebrates collection at Auckland Museum, but I knew nothing of this until I had been at the museum for twenty years.

In the mid 1980s I became aware that there was a large fossil egg in the collection that was partly encased in fine-grained sediment. It was about 208 millimetres long and the shell was ivory-coloured. On one side the shell was exposed. Here the egg looked whole, but was broken into large pieces that were being held together and kept in place by the sediment. The other side of the egg was staved in and obscured by sediment. There was no label with the egg, nor any number written on its surface that could provide a link to information recorded elsewhere.

Natural history specimens that have never been labelled or numbered, or have lost their labels or numbers, are a disappointment and a tragedy, large or small. I had no way of knowing where this egg had been found, or by what circumstances it had come to the museum. Lurking in my mind was the thought it was most likely from New Zealand, and therefore probably the egg of an extinct moa. As such it would be a rare treasure and deserving of extra care and attention.

Large quantities of moa bones, from swamps, caves, sand dunes and Māori kitchen middens, or rubbish dumps, are testament to the former presence in New Zealand of these big wingless birds. Broken eggshell pieces are common but whole eggs are very rare, which is not surprising given the fragility of birds' eggs. In natural history museums around the world, there are only some thirty-six moa eggs that are whole enough to permit their length or width to be measured. Many of these are imperfect, with a large section or several smaller sections missing. Others have been reconstructed—some inaccurately—from broken fragments found together as an isolated group. Moa eggs are thin-shelled. Even the largest have shells only around 1.5 to 1.8 millimetres thick, whereas ostrich eggshell is about two millimetres thick and that of the extinct elephant bird of Madagascar may exceed four millimetres.

I showed the mystery egg to Jack Grant-Mackie, a geologist and palaeontologist at Auckland University. He suggested I send a sample of its sediment to the New Zealand Geological Survey of the Department of Scientific and Industrial Research for petrographic analysis. In 1987 I received a short but detailed report. The size, type and relative proportions of the mineral grains in the sediment were consistent with a New Zealand source, most likely in the eastern South Island, although a foreign origin could not be ruled out.

In the late 1990s I gained funding from the Lottery Grants Board for a project to clean and box the museum's moa bone collection, which was stored on large wooden trays open to dust. As part of this project I was able to employ a conservator, Sue Cooper, to treat the mystery egg and make a special acid-free storage box for it, with a cushioned interior. We decided Sue should remove sediment from inside the egg, leaving a ten- to fifteen-millimetre thickness to hold all the shell fragments in place. This would lighten the egg (it was far too heavy given its fragility), expose whatever lay inside, and make the object look more interesting for future display.

To guide Sue's work in removing the interior sediment we had the egg CT-scanned at a local hospital. Dr Mark Osborne, assisted by Margaret Bowcock, did this for us after-hours and seemed delighted, for a change, to be scanning something other than a part of the human body. I hoped the scan would reveal bones of an embryonic bird inside the egg, but the results were more commonplace: the sediment was simply hiding large pieces of shell that had shifted as the egg had caved in on one side. Most of the egg's outer surface was worn, quite badly, and I hoped the buried shell would be pristine. However, it too was weathered. The egg, abandoned by its parent for whatever reason, must have lain about long enough to be affected by sun, rain and wind-blown sand before one side collapsed and it was buried by sediment, preserving it for discovery.

Sue and I wrote a report on this "probable moa's egg" that was published, with photographs, in *Records of the Auckland Museum* in December 2001. This annual publication is sent on exchange to other institutions in New Zealand and abroad. A copy of the volume in question arrived at the library of Taranaki Museum in New Plymouth and was seen by the director, Ron Lambert. Ron rang me in January 2002. From the photographs, he had recognised the egg as one that had been in the possession of Blanche Stuart Halcombe, the woman who had died in the New Plymouth house fire.

A newspaper photograph from the time of the fire showed a neighbour holding a large egg in a small glazed case. I studied the photograph. A pattern of crack marks on the egg-shell matched that of the egg in Auckland Museum. Blanche Halcombe had bequeathed various parts of her collection of natural history specimens to Auckland and Taranaki Museums. The will and legal papers may be sitting somewhere in Auckland Museum's archive, but when they were received in 1962 it seems that no copies were made, or retained, to accompany the egg into the land vertebrates collection.

From the Halcombe papers at Taranaki Museum, Ron copied me a handwritten letter dated May 27, 1939 from Blanche addressed to "Cousin John". Blanche wrote: "My brother got a moa's egg (dredged up at Cromwell)." This established where the egg was from and agreed with the Geological Survey's prediction of an eastern South Island locality for the associated sediment.

I later discovered that on August 9, 1898, at a meeting of the Otago Institute, there was exhibited "the greater part of the eggshell of a moa … found in sandy soil near Clyde, Central Otago". This egg was deposited at Otago Museum by a Mr Turton, and as Turton was Blanche Halcombe's family name this was likely to have been her brother. Clyde and Cromwell are about twenty kilometres apart and Blanche had perhaps used Cromwell in error in her 1939 letter.

All evidence points to this egg being from a natural alluvial site, not from an archaeological one, and dating from the Late Holocene Epoch. It is at least 500 years old, since all moa were probably extinct after that, but probably no older than a few thousand years. Mark Horrocks, an expert in pollen analysis, took a small sample of the sediment and reported in 2002 that pollen in it was at low concentration, and poorly preserved. However, one pollen grain from an endemic forest tree—rimu, *Dacrydium cupressinum*—reconfirmed that the egg was from New Zealand. In 2008 Leon Huynen, a molecular biologist at Massey University, took a small scraping from the surface of the eggshell, and isolated fragments of DNA that showed the egg had been laid by the large South Island form of *Euryapteryx*, the stout-legged moa.

Blanche Halcombe was born in New Plymouth in 1881. She was a granddaughter of William Swainson, who had been born in London in 1789, emigrated to New Zealand in 1841, and died there in 1855. Swainson was an amateur naturalist and artist who travelled in the Mediterranean, South America and Australia, and published many significant works on both zoology and botany. He was a pioneer in the use of lithography, a technological advance on engraving, to illustrate zoological subjects in books.

Blanche, too, had a passion for natural history and spent much time in the New Zealand bush. She built up a large collection of natural history specimens, both from her own collecting and from among the items brought to New Zealand by her grandfather.

In 1910, at the age of twenty-nine, she moved to England to serve as a lady's companion. There she caught up with a cousin, Norman Halcombe, who had left his job as a civil engineer in San Francisco to sign up with the British Army. Halcombe survived the war and he and Blanche were married in London in January 1919. They had just one night together as a married couple before Norman left for an army posting in India. There was an influenza pandemic at the time and Norman contracted the disease on board the ship. It progressed to pneumonia and he was taken ashore at Port Said in Egypt, where he died in mid February at the 76th Casualty Clearing Hospital. The army chaplain wrote to Blanche:

18 February 1919

Dear Mrs Halcombe,
You have my greatest sympathy over the death of your husband.

I buried him in the Port Said cemetery which is quite a nice one surrounded on three sides by the sea. In a few

> weeks a cross will be put up. All the graves are covered in a green creeper – more of an ice plant than anything, with pretty daisy-like flowers …

Blanche returned to New Plymouth, living frugally on her husband's war pension. In later years she lived alone and her house, which contained her natural history specimens, became cluttered with excess belongings and rubbish. The garden was rampant. As she grew older Blanche had poor health, including spinal arthritis, which limited her mobility. She had little money and eventually the telephone, electricity and water were cut off. She was increasingly reclusive, barricading doors and windows against unwelcome visitors. She corresponded regularly, and there was a clear path in the overgrown garden from the front door to the letterbox, but neither her correspondents nor her neighbours knew how dire her situation was.

Most of the treasures were in the living room, where Blanche slept and cooked on a Primus stove. There was a sea of rubbish piled up to the level of the windowsills. One day neighbours saw smoke coming from the house. The fire brigade had difficulty getting through the garden, breaking into the house and reaching the source of the fire amid all the clutter and squalor. Blanche was still alive but died in the ambulance. It was left to her neighbours and staff from Taranaki Museum to salvage what treasures, including valuable artworks, they could from the damaged house.

The moa egg was among the items rescued, adding a dramatic chapter to a story that had begun many centuries before, when a stout-legged moa had laid the egg near a river in Central Otago. The egg had failed to hatch. It had lain about while its surface weathered, and then it broke and was filled and covered with sand. It remained buried until about 1898, when it was discovered by chance, perhaps by a workman during gold-dredging operations. Blanche's brother obtained the egg and showed it off in Dunedin. It passed to Blanche and was one of the treasures in her private collection until pulled from the fire-damaged house. Remarkably, the egg was unburnt.

The moa egg then came to Auckland Museum, where its significance was overlooked for forty years until its identity and history were finally pieced together and it took its place as one of the few known whole eggs of these remarkable birds. Its rarity gives it great scientific importance and also a large monetary value, perhaps around half a million dollars.

Auckland Museum has two other moa eggs whole enough to be included among the world list of thirty-six whose dimensions can be measured. In March 1931 a man called Louis Griffin spent nearly two weeks at Mangōnui Hotel on the east coast of the North Island's northern tip. Griffin was Auckland Museum's preparator and assistant director. He studied fish and the aims of his visit were to obtain rock-pool fishes for the museum's collection and investigate reports of moa bones in the sand dunes.

Roads in Northland were poorly developed in the 1930s, so Griffin travelled from Auckland to Mangōnui on a boat, SS *Clansman*. Once there he telephoned an elderly man,

Louie Matthews, who farmed at Aurere near the southern end of Tokerau Beach in Doubtless Bay. He then got a car, collected Matthews and drove to Tokerau Beach, where Matthews showed him the moa bone sites. Griffin wrote to the museum director in great excitement, reporting "a veritable necropolis of moas, extending for miles along the inner lines of sand dunes of the bay".

At their home, Louie Matthews and his wife showed Griffin their collection of curios found in the vicinity over many years. Among them were two whole moa eggs. When Griffin emphasised the rarity of these eggs and their biological importance, the Matthews agreed to donate them to the museum collection. In response to a subsequent letter of thanks from the director on behalf of the museum and its governing council, Louie Matthews replied: "It will be a lasting satisfaction to know the eggs we had treasured so carefully for nearly 30 years in our home have now a final resting place, as additional objects of interest in your splendid museum."

Nobody at the Mangōnui Hotel knew of the treasures in Griffin's room. He packed the eggs carefully in a small tin case, which travelled on his knees during much of the return journey to Auckland. Both eggs had been found around 1900 in the sand dunes at the southern end of Tokerau Beach. One egg, measuring 120 by 91 millimetres, remains almost perfect, with just a small hole near one end, several minute holes, and some cracks. The other, 121 by 97 millimetres, has a chunk, representing about a fifth of the shell, missing at one end. This egg was found first, at an eroding stream-bank in the dunes, the near perfect egg a little later. The eggs are greyish-white in colour but the outer surfaces are badly worn and in many places are chalky and pitted.

While covered by sand, animal remains in the northern sand dunes are protected, but when wind and rain erode the dunes and the remains are exposed they deteriorate in the harsh sunlight and under the abrasive action of wind-blown sand. They are also at risk of being trampled by wandering stock—and these days from people and all-terrain vehicles. Eventually, all exposed bone and eggshell will turn to dust in the harsh environment. The Matthews had found the two precious eggs just in time to save them from certain destruction.

The Tokerau Beach eggs are the smallest known moa eggs and probably belonged to the coastal moa, the small northern form of *Euryapteryx*, whose bones predominate among those that have been found in the area. The better of the two eggs, being almost perfect, might have an insurance valuation of around a million dollars.

In 1972 the museum's preparator, Leo Cappel, made casts of this egg so a copy could be displayed at the museum in Kaitāia, near where the egg had been found. Making the initial mould must have been nerve-wracking. A second cast was retained.

In the late 1990s, when the exhibits in Auckland Museum's natural history galleries were revamped, a large case was assigned for a moa display. At the back of the case there were several large articulated skeletons, with smaller items positioned at the front. I wanted to display the largest and smallest moa eggs to show their relative sizes. To represent the largest we already had a cast, and for the smallest I decided to use the 1972 cast instead of the real Tokerau Beach egg: it was too dangerous to put such a small, fragile and unique item in a big case with larger objects.

In 2014 a pane of glass in the roof of the moa case collapsed inwards, skittling both moa egg casts. It was a close call. The smaller model received surface gouges, which one of the museum's conservators repaired, but it would have been curtains for the real egg had it been in the case. Canterbury Museum was not so lucky—in 1990 its best-preserved whole moa egg was accidentally broken during casting.

The man who imagined the moa

Auckland Museum's land vertebrates collection was organised quite well before I took charge of it but major problems remained. The numbering system for specimens, dating from 1925, was an unworkable species-based system that tried to incorporate taxonomy into the registration number—a futile exercise and an unnecessary complication. The system needed to be replaced by simple consecutive numbers so specimens would not have to be renumbered every time they were re-identified as something different, or when species were merged or split.

As well as this, every specimen needed to be retrieved and checked, because important information on accompanying labels had not always been transferred to the catalogue cards, which were the principal record of the collection. And there were many specimens that had not yet been numbered and catalogued at all. I took a deep breath and embarked on a decades-long project.

Sometime in the 1990s, as I worked gradually through the revamp with a little help from technical staff and volunteers, I came to the moa eggshell fragments. Moa, New Zealand's large wingless birds related to the kiwi, emu and ostrich, were quickly wiped out after Polynesian settlement. Natural history museums have a unique role as repositories of the remains of extinct animals and plants. In New Zealand this includes guardianship of large collections of moa bones—pretty much all that remains of these intriguing birds. In fact there are a few other vestiges of moa, including some whole or almost whole eggs (see pages 40–51) and shards from broken moa eggshells. Eggshell fragments have been found in large quantities at older archaeological sites around the country, and at many natural fossil sites, most younger than the last Ice Age that ended about 10,000 years ago.

Of the many groups of moa eggshell fragments at Auckland Museum, two assemblages stood out, one very large and both very old. The labels, in faded handwriting on yellowed paper, showed the fragments had been collected by a W.B.D. Mantell. A group of about forty were from Ōhawe on the Taranaki coast of the western North Island, and had been collected in 1847 or 1848. The second assemblage comprised some 1,600 shards from an archaeological site at Ōamaru, in the eastern South Island, collected in 1852. Auckland Museum was just being founded in 1852 but one of the labels is dated April 15, 1890, probably the date the museum received the specimens. A nice touch is that three of the Ōhawe shards have "Walter Mantell" or "W. Mantell" written on the inside egg surface in small copperplate writing, presumably by the man himself.

Walter Baldock Durrant Mantell was born at Lewes in Sussex, England in 1820. He was training for medicine when in 1839 he abruptly chucked it in and departed for

New Zealand, arriving in Wellington at the end of January 1840. He held various administrative jobs that culminated in his appointment as a Commissioner of Crown Lands in 1851. Mantell became disillusioned with the government's acquisition of Māori land for European settlement, and in particular with the failure to set aside other than token reserves for the Māori population in some regions. He turned to politics and was an elected parliamentarian from 1861 to 1866. He then took up a seat in the higher chamber, the Legislative Council, which he held until his death in 1895.

This story is convoluted, and before we stray too far from moa eggshell fragments I should say a word about my own research on that subject. Despite the abundance of broken moa eggshell, only the microscopic structure and chemical composition had been studied in much detail. There had been little research on what the presence of eggshell could tell us about moa in the area. Reports of archaeological and palaeontological excavations typically list "moa eggshell" without giving any further analysis.

It had been known since the 1930s that moa eggshell fragments were either "thick" (about 1.3 to 1.7 millimetres) or "thin" (about 0.5 to 1.0 millimetres). These are not big differences yet by taking a fragment between your fingers you can generally feel whether it is thick or thin, a testament to the remarkable sensitivity of human fingers.

Thicker eggshell fragments likely come from larger eggs produced by the larger species of moa. To test this I tracked down whole moa eggs in museum collections. I found thirty-six whose length or width could be measured but they varied greatly in their degree of wholeness (some being half missing) and reliability (some being inaccurately reconstructed from fragments). The damaged ones were of great benefit because at the broken edges I could measure the thickness of the shell using simple instruments—vernier calipers or a screw micrometer—and relate it to egg length. For my biggest sample, fourteen South Island eggs, there was a weak but positive correlation between egg length and shell thickness; in other words as one measurement increased so did the other. This strengthened my basic premise that the thicker the fragment the larger the kind of moa it came from.

I then studied museum samples comprising large collections of eggshell fragments from specific localities. I measured as many unweathered fragments from each site as I could, assessing the thickness of each fragment just once, to the nearest 0.01 millimetre. I graphed these thicknesses to obtain a frequency chart for each site that was like a profile diagram. We know from the distribution of moa bones that in any given part of New Zealand there were up to five or six moa species but usually fewer. My hope was that the distribution of thicknesses of the moa eggshell fragments at a location would roughly reflect the sizes and relative abundances of the resident species of moa.

Comparing the eggshell thickness diagrams at specific sites with what was already known about the moa fauna from bones at the sites gave a good match, especially for the North Island where there had been fewer types of moa. I included a few archaeological sites—kitchen middens believed to date to one to two hundred years after AD 1250 to 1300, when Polynesian settlement of New Zealand began. Samples from these showed just as large a range of moa eggshell thicknesses as natural fossil sites. This suggested that the Polynesian settlers ate eggs from all available species of moa, and not just the largest ones. This is hardly surprising but nonetheless a line of objective evidence in the reconstruction of the past.

Moa must have been a formidable presence in the lives of the early Polynesian settlers, but the birds were quite quickly exterminated and their details largely lost to memory. Henry Thomas Hill, reviewing knowledge of the bird in 1914, noted that "legends are few, Native information uncertain and unreliable". It was Richard Owen, later to be founding director of the British Museum (Natural History), now the Natural History Museum, who rediscovered the moa for science. The leading British comparative anatomist of his day, Owen in 1839 received a broken shaft of a large thigh bone from Poverty Bay on the North Island's east coast. By comparison with museum reference specimens he determined that the bone belonged not to a cow-sized mammal, as seemed likely from a cursory examination, but to a large bird. At a meeting of the Zoological Society of London on November 12, 1839, the records reveal, he "exhibited the bone of an unknown struthious [ostrich-like] bird of large size, presumed to be extinct ... [and concluded] so far as my skill in interpreting an osseous fragment may be credited,

I am willing to risk the reputation for it on the statement that there has existed, if there does not now exist, in New Zealand, a Struthious bird nearly, if not quite, equal in size to the ostrich."

With this announcement—based on one six-inch bone fragment—Owen predicted that New Zealand had been home to giant birds we now call moa. In those days the proceedings of learned scientific societies were reported in the London press, and relayed around the world to a public fascinated by the progress of science. A gigantic bird in New Zealand was one of many deductions and discoveries that marked out Owen as a brilliant scientist and thrilled the many sectors of society, from the royal family to the general public, who delighted to hear of the novel animals and plants being discovered in the far corners of the world.

Some felt Owen had gone too far with this one, but when entire moa skeletons later arrived in London Owen's conclusion was proved correct. With each consignment Owen wrote more papers on moa, describing new species and recording ever more aspects of their skeletal anatomy. Nine or ten species of moa are now recognised and all but one bear names that Owen bestowed between 1843 and 1870.

After news of Owen's initial announcement, many settlers in the colony were on the lookout for moa bones. Walter Mantell's early work with government agencies took him around New Zealand and he remained alert for the natural history objects that, through his father Gideon, an amateur geologist, had been so prominent in his childhood. During this early period he collected many moa bones and much moa eggshell, including the shell fragments that later came to Auckland Museum.

He sent collections to his father in England, who invited Richard Owen to study them. Among the bones were those of the extinct North Island takahē, a large flightless rail or woodhen. In 1848 Owen gave the North Island takahē the name *Notornis mantelli* in Walter Mantell's honour.

Walter Mantell also obtained the first two skins of the South Island takahē: the birds had been caught and eaten on the west coast of Fiordland, one by sealers in 1849 and the other by Māori soon afterwards. Their skins were sent to the natural history museum in London. The ornithologist Walter Buller noted that they "now stand side by side in the National Collection of Great Britain and, like the remains of the Dodo in the adjoining case, daily attract the attention of thousands of eager visitors".

In 1852 Adam Bartlett, superintendent of London Zoo, named the North Island brown kiwi for Walter Mantell. In his description of the kiwi in *Proceedings of the Zoological Society of London*, Bartlett acknowledged a kiwi specimen (of a different species) that had been passed to him by Walter's father and collected by Walter in Dusky Bay. He concluded: "I therefore propose the name of Apteryx Mantelli for [the] smaller and more common species—a humble effort to commemorate the exertions of Walter Mantell, Esq., to whom we are indebted for so many valuable discoveries in the natural history of New Zealand."

Richard Owen's numerous important papers on moa and other New Zealand birds were republished by him in 1879 as a book entitled *Memoirs on the Extinct Wingless Birds of New Zealand*. The text fills one large volume and the plates, mostly large and exquisitely detailed lithographs, another. The Auckland Museum library has three copies of the book, and the head librarian allowed me to have one precious copy assigned to my office. In places there are notes in the margin of this particular copy, written in ink in a Victorian hand. The notes are mostly disparaging, including exclamations such as "Nonsense" and "Impossible & untrue". Where Owen refers to portions of a skull that was among bones collected by Walter Mantell and sent to London, there is the comment in the margin: "I sent a complete skull." This copy of the book was presented to the library by an Auckland philanthropist, Edward Earle Vaile, but the original owner was surely Walter Mantell himself.

Books about Richard Owen sometimes state he was buried at Ham, then a village south-west of London and now a suburb of Greater London, but I had never seen published any exact details. Surely the whereabouts of the grave of such a luminary would be widely known? I emailed the Royal College of Surgeons, where Owen began his career, and the Natural History Museum but nobody could say exactly where Owen's grave was. In 2007 I looked at maps of Ham on the internet. One road bordering Ham Common was called Church Road and unsurprisingly the maps showed a church and graveyard. This looked promising, and since I was visiting London that year I decided to visit the graveyard.

I took the tube to Richmond and had a short train journey from there to Teddington. I had marked the most direct walking route to Ham Common with a highlighter on a printout of the internet map. Knowing nothing of the area, I was slightly apprehensive that I might be traversing a bleak part of suburbia with prostitutes on street corners and used syringes in the gutters but instead it was a lovely walk. After sitting and eating sandwiches at the pretty Teddington Lock I found St Andrew's Church next to Ham Common. It was a picture postcard English scene with the brick church, mature green shade trees and sunny skies.

The churchyard gate was unlocked. I began randomly pacing between the graves and within ten minutes my pilgrimage was complete. I found myself at Richard Owen's final resting place. The grave was in good order but the adjoining grass and weeds were long. It was clearly not a revered site in the same league as Oscar Wilde's grave in Père Lachaise Cemetery in Paris. Owen's reputation for bad professional behaviour and his rejection of natural selection—he disavowed Darwin's theory—seems to have outweighed for many people his achievements for science and museums. The grave was a small, low sarcophagus faced with red granite, and with inscriptions on the sloping panels that formed the top of the tomb. The inscriptions made clear that Owen's wife and sister were buried in the same plot.

My hunch on the location of the grave had been correct and I felt a need to share my joy. The church and the road were deserted. The only nearby building was the vicarage. I knocked on the door and a minister opened it. He seemed a little put out at the unscheduled interruption and kept me on the threshold. As my story tumbled out about moa, New Zealand and Sir Richard Owen I sensed, perhaps unfairly, that he might be attributing my intrusion to some form of rude colonial behaviour. He hadn't heard of Sir Richard and there was no further memorial to him inside the church. Outsiders visiting the graveyard were usually pursuing a forebear of Queen Elizabeth the Queen Mother.

Although it may not show much interest in Owen's grave, the Natural History Museum in South Kensington does display a full-sized bronze statue of Owen. The great osteologist is holding in his left hand the femur of a moa, an allusion to New Zealand and the famous bone he studied. For a long time this statue occupied pride of place at the top of the stairs in the Central Hall but, with the bicentenary of Charles Darwin's birth in 2009, the museum's statue of Darwin was moved to prime position and Owen's relegated to a side alcove.

When the natural history galleries at Auckland Museum were redeveloped in the late 1990s I proposed that a large case should show our three-metre-tall articulated skeleton of a giant moa—the bones are actually plaster casts—and a life-sized plaster bust of Owen that we had in our applied arts collection. The moa exhibits are a big drawcard for visitors to Auckland Museum and much photographed. In 2006 a second edition was published of a textbook on dinosaurs by an American palaeontologist, Anthony Martin. In a chapter on birds as modern dinosaurs there was a photograph, taken by a museum visitor, showing our display of the moa skeleton and bust of Owen.

In Britain the illustration was seen by Ray Winder, a

wood sculptor and artist who was researching the portrait busts and paintings of Owen. In 2010 Winder contacted the Auckland Museum library seeking further information on the museum's bust of Owen. The enquiry came to me to answer, with interesting results. All we knew was that the bust had been received by the museum in 1932 from Archbishop Alfred Averill. Born in England in 1865, Averill had come to New Zealand in 1894 after serving in several London parishes. He returned many times to Britain, including to attend Lambeth Conferences, and this was probably when and where he had obtained the bust.

Ray Winder sent me a copy of a photographic illustration that had appeared in a 1946 book about the Royal College of Surgeons museum in London. It showed a bust of Richard Owen as a young man by a sculptor, Edward Hodges Baily, who also produced the figure at the top of Nelson's Column in Trafalgar Square. The bust was clearly the same as the one at Auckland Museum. Ray speculated that the museum's plaster bust may have been an intermediate stage in the production of the final version, which was carved in marble. In 1941 the marble bust was destroyed when the Royal College of Surgeons building was hit by an incendiary bomb dropped by the German Luftwaffe. The plaster version at Auckland Museum is likely to be the sole survivor of this work of art, the earliest portrait bust of Owen.

Besides their known and celebrated objects, museum collections also harbour hidden gems—objects whose true significance awaits discovery. The story of this plaster bust shows the sort of unlikely sequence of serendipitous events that can result in recognition of a major museum treasure.

Flight of the long-tailed cuckoo

In March 1999 I did a post-mortem examination of a long-tailed cuckoo, *Eudynamys taitensis*, that had died accidentally, and been given to the museum by the Department of Conservation. It had been kept frozen, so after thawing it I weighed it and took a series of standard measurements. It was a handsome bird, with different shades of brown in streaks and bars on its plumage.

Long-tailed cuckoos weigh about 125 grams, similar to the weight of a common myna, *Acridotheres tristis*, but they have a very long tail—in this individual it was 235 millimetres. They also have the pointed wings that typify a strong long-distance flier. And then there are the toes. A curiosity of cuckoos is that they have a "zygodactylous" toe arrangement like parrots: two toes point forwards and two backwards. Birds typically have four toes, having lost the fifth one, the equivalent of our little toe. In most, three toes point forwards and one points backwards, the "anisodactylous" condition.

My next step was to open the abdominal cavity so as to remove the stomach (for later analysis of the bird's diet) and to determine the bird's sex by examination of the gonads. In birds, unlike in mammals, the testes are internal. The female sex organs are reduced to just one set (as a rule) on the left side only. This oddity is an evolutionary adaptation to reduce weight for the sake of flight.

The belly of E39—it was the thirty-ninth long-tailed cuckoo I had dissected—had already been ripped open, perhaps by a predator or during a fatal collision. When I cut the oesophagus above the stomach, and gently lifted out the stomach with the rest of the intestinal tract attached, there was a surprise underneath. The bird was a female, with a grossly enlarged, fleshy and convoluted oviduct that looked for all the world like a second intestine. I felt something hard and realised there was a shelled egg inside the oviduct. The bird had died at that rare moment when an egg was ready to be laid.

During formation of a bird's egg the largest developing ovum is first released from the ovary. Instead of wandering into the body cavity, it is caught by the end of the oviduct—the infundibulum or "funnel", which is shaped like a baseball glove—and begins a passage inside this fleshy tube. If at least one spermatozoon has swum this far following a recent copulation, fertilisation takes place. Then, in a particular region of the oviduct, the ovum is coated with albumen, the egg white. In another region, as it passes down, it receives its shell membranes.

After this the hard shell is laid down, beautifully shaped by the egg's rotation at that point in the oviduct. Finally, any shell pigments are deposited to give the egg its spots or streaks. E39 had died when her egg was at this point. The egg was cracked, but after I flushed away its contents and thoroughly dried it I was able to partially reconstruct and strengthen it by using as glue an acrylic plastic, Paraloid B-72, dissolved in acetone.

The long-tailed cuckoo has speckled eggs that mimic the eggs of the host birds it parasitises—whiteheads in the North Island and yellowheads and brown creepers in the South. After the female cuckoo has removed a host egg and added her own to the host's clutch, the two kinds of eggs are almost indistinguishable.

The host birds nest in forest and scrub. Because their nests are hard to find, the breeding of the long-tailed cuckoo has not yet been studied closely. There are just a handful of long-tailed cuckoo eggs in museum collections and some are suspect since they have been taken from nests and merely assumed to be cuckoo eggs: some may be host eggs. The egg from E39 was not just cracked but pale and anaemic, suggesting there had been insufficient time for the pigments to be fully applied to the shell. Nevertheless it is a treasure because eggs of this species are so rare and this one's identity is certain.

The long-tailed cuckoo, koekoeā, is seen in New Zealand during summer but almost never in winter. I looked at what was recorded of Māori knowledge of the bird, starting with Margaret Orbell's *Birds of Aotearoa: A Natural and Cultural History*. Orbell's summary was based on early accounts by missionaries and ethnologists who, during their travels, had recorded what Māori told them. Writers of ancient Greek texts and the Bible made reference to seasonal movements of birds, yet in Europe until the late eighteenth century the idea persisted that hibernation might be the reason many birds were absent in winter. Similarly, there seems to have been no Māori awareness of the long-tailed cuckoo as a seasonal long-distance migrant: many believed that during winter the birds hid in mud, among stones or in holes in trees. Others thought the birds' appearance in spring could be attributed to geckos or tuatara periodically going back to the water that was once their home, entering cocoons, and emerging later as koekoeā.

To post-modernists and cultural relativists in the humanities, these statements about hibernating European birds and the transformation of New Zealand reptiles into cuckoos will be true, and as valid as any scientific assertions on the subject. However, for me these colourful explanations exemplify pre-Enlightenment commentary on the natural world, formed without intellectual mechanisms with which to pursue objectivity. The Royal Society of London, one of the world's oldest surviving scientific societies, traces its origins to 1660 and was shaped by Enlightenment thinking. Its motto, a catchcry for science, is *Nullius in verba*—Trust nobody's words. In line with this, and benefiting from modern circumstances, my task was to travel to museums around the world to examine long-tailed cuckoo specimens for an objective look at how the birds' distribution changed with season. Science is hard work compared to relying on hearsay and guesses.

Members of Captain James Cook's first antipodean voyage in 1769 encountered the long-tailed cuckoo in New Zealand. Based on material obtained from French Polynesia in 1773 during Cook's second voyage, the bird was described and named in 1787. In 1843, in his book *Travels in New Zealand*, German naturalist Ernst Dieffenbach stated that the cuckoo was migratory, "appearing near the coasts [of New Zealand] in the month of December". In 1844 British ornithologist George Gray reported in *The Zoology of the Voyage of H.M.S. Erebus and Terror* that the cuckoo "comes from the north to the neighbourhood of Port Nicholson [Wellington] in the month of October, and returns in April".

However, long-tailed cuckoos were sometimes recorded on Pacific islands during the breeding season, and so were considered by some to be year-round residents of Polynesia. Moreover, a long trans-Pacific migration over water was questioned. Alfred Russel Wallace, co-proponent with Charles Darwin of the theory of evolution by natural selection, in his 1876 textbook of biogeography considered the distances from New Zealand to Pacific islands made such journeys extremely improbable. He felt it was premature in so sparsely settled a country to rule out a purely internal New Zealand migration.

Then in the 1920s the Whitney South Sea Expedition systematically collected birds on islands across the tropical South Pacific. Back in New York, Cardine Bogert at the American Museum of Natural History examined the long-tailed cuckoos that had been collected, and other museum specimens of the bird, and in 1937 published a paper suggesting that long-tailed cuckoos from New Zealand and the tropical Pacific were the same population.

This has not been disputed. The birds breed only in New Zealand and have a vast wintering range in the Pacific. The migration is spectacular considering that such a small land bird embarks on such long journeys over water. Only a handful of long-tailed cuckoos have ever been banded with numbered rings in New Zealand and none has been recovered overseas, so precise details of the migratory flights are lacking.

My own study was a repeat of Bogert's, which indirectly examined the pattern of the cuckoo's migration. There are now many more museum specimens of long-tailed cuckoos than in 1937, and many more published sightings. Each properly documented museum specimen or sighting is proof that the cuckoo lived at a certain place in a certain month. I got about 570 records of long-tailed cuckoos in New Zealand by combing through annual summaries of bird sightings published from 1943 to 2006 in *Notornis*, the journal of the Ornithological Society. I collected some 170 sight records from scores of books and periodical articles on the local ornithology of the various Pacific island groups, many of these materials conveniently at hand in the Auckland Museum library. From Auckland Museum, and by visiting New Zealand's three other main museums, I got data for most of my sample of 180 cuckoos collected in New Zealand. Data for cuckoo specimens from Pacific islands—200 specimens—were mostly obtained when I visited the natural history museums at Paris, London (the national bird collection is at a satellite location at Tring, Hertfordshire), New York and Washington, D.C.

The American Museum of Natural History in New York has the largest collection of long-tailed cuckoos: it took me all day in a basement storeroom to go through them, just noting a few simple details. This holding results both from the Whitney South Sea Expedition and the museum's purchase in 1932 of the private bird collection of Walter Rothschild. Rothschild's 280,000 study-skins from around the world made up the largest private collection of stuffed birds in the world. He was forced to sell it to ease a financial situation made dire by a blackmailing mistress. With the help of the Whitneys, the bulk of the collection was purchased for US$225,000 and shipped from England to New York.

Curators and collection managers at seven other museums, from Philadelphia to Berlin, sent me details of their long-tailed cuckoo specimens, often with photographs. Altogether I amassed some 1,120 records, many of them new since Bogert's study. These I mapped month by month to illuminate the fine details of the cuckoo's migration between the temperate south and tropical north.

The museum specimens allowed consideration for the first time of migration patterns in relation to age. Long-tailed cuckoos are difficult to observe, and published records are usually of birds seen only fleetingly. It's rare for an observer to see the plumage clearly enough to assign an age. However, I was able to check the plumage of each museum specimen and age the bird. Adult and immature birds have very different plumage. The back of the adult bird is dominated by transverse reddish-brown bars; in an immature bird large pale cream-coloured spots dominate. The adult has white plumage between the dark longitudinal streaks of its under parts; in the immature bird this is pale brown.

Results confirmed a vast fan-shaped distribution of the long-tailed cuckoo, extending 6,000 kilometres from New Zealand north to the tropical Pacific, and 11,000 kilometres from east to west in the tropics. Wake Island in the north (19.3° N), Palau in the west (134.5° E), Henderson Island of the Pitcairn group in the east (128.3° W) and the Snares Islands in the south (48.0° S) are the extreme records in this range. The distances from central New Zealand (say, Nelson) are 5,680 kilometres to Henderson, 6,640 to Palau and 6,700 to Wake.

The cuckoos are virtually absent from New Zealand from June through to August, and rare in May and September. During these months they are numerous in the Pacific islands. Museum specimens revealed that almost all long-tailed cuckoos returning to New Zealand in October are in adult plumage, meaning even the youngest returning birds have by then lost their immature plumage. Cuckoos with immature plumage are almost entirely absent in New Zealand until late summer and autumn, so they must be hatched that summer and can be no older than a few months at this time.

In the northwards migration, dates on museum specimens showed that adults move first, with immature birds lagging up to three months behind. This means that most young long-tailed cuckoos cannot get clues from their parents on where to fly in the wide Pacific: we don't understand how they choose a direction and destination.

The return journey south to New Zealand, when the young birds are ready to breed, is less of a wonder. They probably follow a homing instinct to head back to their place of hatching. After they hatched they would have learned (by "imprinting") the characteristics of the host species that raised them, and to perpetuate the host-parasite breeding system they simply have to seek out the same species as host for their own breeding attempts.

Satellite transmitters for attachment to wild animals are now small enough to fit birds the size of the long-tailed cuckoo. In a study based at Massey University's Auckland campus, zoologist Michael Anderson attached transmitters to cuckoos over two summers from 2013. The transmitters were expensive, as were the data fees from the company that operated the satellites. The cuckoos were very hard to catch. After much effort and expense, Anderson managed to release a handful of birds with transmitters. This has been rewarded by some detailed and fascinating insights into the cuckoos' island-hopping as they move to and from the tropics.

It has also shown that the birds are such strong and competent fliers they may make frequent local movements in New Zealand—for example, between Little Barrier Island and the mainland. This solved a mystery. In the Auckland Museum collection there are some long-tailed cuckoos that were found dead in the city area around November. This is the cuckoo's peak egg-laying month, and a time when they should all be in their breeding locations. There are no host birds for them in Auckland so they cannot breed there. The nearest parasitised whitehead population is on Little Barrier Island, 100 kilometres away. Now we can surmise that a cuckoo in Auckland in November is on a day excursion from a distant breeding site.

During my years as curator we received between fifty and a hundred dead birds of all kinds from members of the public every year, and there were always two to three hundred in the freezer awaiting preparation. Being particularly interested in cuckoos I examined each one that came in to record useful data before the bird was processed into a stable specimen such as a study-skin or bones. Long-tailed cuckoos arrived at an average rate of 2.5 birds a year, and I handled about seventy-five in total. Only recently have I begun the task of opening the stomachs, which I had preserved in alcohol, to identify and quantify the contents for an analysis of diet.

Long-tailed cuckoos have a thick powerful beak about two and a half centimetres long. The contents of each stomach are a tantalising surprise as you cut open the muscular gizzard wall and tease out the remains of the cuckoo's recent meal. To date, the stomachs have revealed mostly large insects, such as cicadas, stick insects (phasmids) and wētā (wingless crickets). But several stomachs have given me a start. Some amorphous blobs have proved to be remains of baby birds—either young nestlings or, in some cases where eggshell was present, perhaps full-term embryos. It's not easy to identify their species and I hope DNA tests will establish whether the cuckoo had pillaged a nest of one of its host species or carried out random predation on an unconnected species.

Cuckoos, with their spectacular migrations and surprising brood-parasitic habits, are hugely fascinating. It has been a pleasure to use collections in natural history museums to uncover some of their secrets.

Booby eggs and a solar eclipse

John McLean was a sheep farmer from Poverty Bay on the east coast. He accompanied New Zealand's great nature writer Herbert Guthrie-Smith on bird expeditions, and in his 1914 book *Mutton Birds and Other Birds* Guthrie-Smith described him as "a friend of older standing … gifted with the most imperturbable of tempers". On a trip to Stewart Island things had become dire. Waist-deep in water chilled with melted hail, and with a rising river to swim, McLean still took time to confirm a pair of rare birds, South Island kōkako, in the flooded scrub. Guthrie-Smith noted: "I acknowledge he beat me there. If I had seen a moa I should have let it pass …"

Geoffrey Buddle, an engineer and ornithologist from Auckland, was also tough. With the Royal Engineers in France during the Great War, he helped build a bridge while under fire, for which he was awarded the Distinguished Service Order. Then he was badly gassed and fought a long struggle to regain his health.

McLean was sixteen years older than Buddle. Whether the two men ever met or corresponded I do not know, but they had much in common besides grit: they shared a passion for natural history, collected birds' eggs and were pioneering nature photographers. The egg collections of both men, and many of their photographs, came to Auckland Museum. McLean and Buddle carefully marked their eggs with identifying codes, and recorded where and when they had been collected. This was critical in giving these objects an importance now that is both scientific and historical.

Eggs, or rather their empty shells, are a major element in museum bird collections. All birds lay hard-shelled but porous eggs: it is one of the main avian characteristics. The shell is made of crystalline calcium carbonate in a protein matrix and is a marvel of engineering—strong yet light.

It seems that birds have no alternative but to lay eggs. They cannot retain developing embryos within their body and give birth to live young as most mammals do because it would make them too heavy. Birds' bodies are scaled down in every possible way to reduce weight for successful flight. Interestingly, flightless birds still lay eggs: perhaps it is now too complex and entrenched a characteristic to be easily changed. Evolution doesn't need to produce the best or ultimate solution to any situation, only one that works.

Birds' eggs are easy to gather, prepare for museum collections, and store. Their size, shape, colour, pattern, microstructure and chemical composition, along with clutch-size—the number laid at a single time—can illuminate much about the biology of birds. Large egg collections give researchers a time series of specimens. Collections like this were used to show that eggshell thinning in wild birds could be correlated with contamination of the environment by the insecticide DDT. This was among the evidence that led to the almost worldwide ban on widespread use of DDT in agriculture from the 1970s.

To clean an egg for a collection you might think it logical to make a small hole at each end and blow out the liquid contents. I remember doing this as a child under guidance from my father. However, the professional method is to empty the egg through a single hole in one side. You prick a small starter hole and then use a metal drill bit that tapers to a point and has fine ridges on the working face. By placing this in the starter hole and rotating it rapidly between the fingers you can grind out a neat round hole of whatever size required, lubricated by the escaping egg white.

The egg can then be held over a sink with the hole pointing down. A curved tube is inserted into the hole and gently blown. As air enters the egg it forces the contents out the hole. Then it is an easy matter to run water in and out of the egg to clean it. Once the egg is dry, the museum registration number can be written next to the hole. The single hole at the side ensures there is no damage to the ends of the egg. This is important because the most common measurement of any egg is its length. The other benefit of a single hole is that when the egg is placed hole-side down for exhibition, it looks intact.

Egg collecting was a common hobby for both children and adults from Victorian times to the middle of the twentieth century. The study of eggs is called oology. For serious collectors in Europe and North America there were oological clubs to join. Magazines were produced and certain companies stocked the relevant equipment.

Then passions changed. Local wild birds and their eggs became legally protected in most Western countries and egg collecting died out, except for the special case of natural history museums, which add to their egg collections as part of their task of making a permanent and verifiable scientific record of biological diversity and its distribution geographically and through time. Many large and well-documented private egg collections have now passed to public museums and become treasured items. The collections of John McLean (380 clutches) and Geoffrey Buddle (460 clutches) are vital elements in Auckland Museum's egg collection of 3,500 clutches or single eggs. An Australian example of a private collection now in public hands is the H.L. White collection of 4,000 clutches at Melbourne's Museum Victoria.

In my time at Auckland Museum we added many new eggs to the collection, including those of rare species such as New Zealand dotterel, North Island saddleback and North Island kōkako. All of these were salvaged from abandoned nests, unhatched eggs in nests and so on, and were usually single eggs or part-clutches, brought in by diligent researchers and conservation workers who understood the importance of natural history collections, or by members of the public.

Results were variable. As often as not the egg might already be weakened by cracks, or it might contain a well-developed embryo and so could not be blown. Sometimes the contents were rotten and had to be removed in a laboratory fume cupboard. The McLean and Buddle eggs are of a higher standard because these collectors sought whole clutches, often observing a nest until all the eggs had been laid. And where possible they took eggs early in the incubation period to ensure their contents were liquid.

John Chambers McLean was born in Hastings, New Zealand in 1871. After his secondary schooling he managed Waikohu Station, his family's large sheep farm at Te Karaka in the hills north-west of Gisborne. He also supervised bush clearing further inland at his hill-country run on the slopes of Mount Maungahaumi. All this brought him in close contact with birds. Between periods of farm work he found time for observations, copiously filling notebooks with his neat writing. He also undertook nature photography, which was difficult and pioneering given the primitive equipment of the time.

McLean collected birds' eggs from 1881 to 1918—the earliest, an egg of a red-billed gull from Cape Kidnappers, when he was just nine or ten. He corresponded with a Charles Holdsworth of Cheshire in England, who probably set him up for egg-collecting by sending him an egg-blowing tube in 1884. McLean's register of eggs, arranged according to bird family on large printed loose-leaf ledger sheets, shows the set-marks—the codes written on the eggs—and collecting details. Most of the eggs came from within thirty kilometres of the home farm at Waikohu. He also often collected south towards Napier, with repeated visits to Cape Kidnappers, a spectacular clifftop gannet colony, with shags, gulls and terns also breeding nearby, and at the braided riverbed of the Tukituki River, where he found eggs of shags, stilts, gulls and terns. Altogether, McLean's collection contains eggs of fifty-seven New Zealand species, and includes important breeding records for birds of the eastern North Island. There are also twenty samples of feathers, mainly from the linings of duck nests, and one of sheep's wool.

Between 1889 and 1912, McLean published eleven articles on New Zealand birds in the British ornithological journal *Ibis* or, from 1902, in a new Australian journal, *Emu*. (A local version, *New Zealand Bird Notes*, later called *Notornis*, was not launched until 1942.) Although McLean was an active collector, his notebooks give the impression that at times he took nests only where tree felling was underway. In his notebooks and publications he expressed concern at the decline in native birds, some of it due to his own bush clearing. He was pessimistic about the future survival of some species, noting the detrimental effect of loss of habitat and the severe influence of introduced predators, particularly rats. He was, though, optimistic that bush birds might survive in areas of high-country forest that had been less modified.

McLean exchanged letters with many notable ornithologists, including Walter Buller in New Zealand, Otto Finsch in Germany, Philip Sclater in Britain, and Australia's Archibald Campbell, whose influential book *Nests and Eggs of Australian Birds* was published in 1900. The foreign eggs in McLean's collection likely came from some of these correspondents. There are Pacific eggs from two small islands in the Phoenix group: Hull (now part of Kiribati) and Howland (now a United States possession). Many of the Australian eggs are from two small tropical islands: Rocky in the Gulf of Carpentaria and Raine on the northern Great Barrier Reef.

The eggs from all these islands are mostly of boobies (tropical gannets), noddies (tropical terns), tropicbirds and frigatebirds. They are all marked with collector names or initials from one family, the Ellises. Albert Ellis was a New Zealander who wrote a book, *Adventuring in Coral Seas*, published in Australia in 1937. Members of the family worked on islands of the Phoenix group, and on Rocky and Raine Islands, mining guano or establishing copra plantations. The eggs are valuable early records of breeding colonies of birds on these islands.

The booby eggs from the Phoenix Islands were collected between 1888 and 1890. Booby eggs are five to six centimetres in length and long and narrow. The underlying shell is pale blue, but both booby and gannet eggs are typically covered by a chalky layer, making them appear white. The chalky surface gets dirtier and more scratched the longer the egg is in the nest. Gannets and boobies breed in large, often dense, colonies on coastal cliffs and on rock stacks and other islands, including low sandy ones, where they are fairly safe from land predators. Both sexes incubate the eggs, but as the birds have no brood-patch on their belly the eggs are incubated under the webs of the feet.

John McLean was a keen observer of nature in a remote part of New Zealand during the early stages of European modification of the landscape. The country had few natural historians, and his egg collection and photographs, together with his published papers, are a significant early contribution to New Zealand ornithology. In 1918 his life was cut short at forty-six: he attempted to ford the flooded Waipāoa River at a crossing near his home but was swept off his horse, and his body was found about two and a half kilometres downstream. This was the fate of many who lived and worked in isolated areas and daily had to cross waterways that could easily become treacherous. It came to be called "the New Zealand death".

Geoffrey Armstrong Buddle grew up in Auckland and attended Auckland Grammar School. His parents encouraged an early interest in photography, and a visit to Little Barrier Island in 1904 seemed to turn his attention to photographing and studying nature. The earliest egg he collected was that of a white-faced storm petrel on the Mokohinau Islands when he was fifteen.

As a young man he studied engineering at the University of California, Berkeley, and at the outbreak of the First World War he was working in Britain. He joined the Royal Engineers and served at Gallipoli until that campaign ended. In France he was at the Somme and the battle of Arras. After the war he joined a firm building docks in Calcutta but the gassing had damaged his health and he had to be invalided to a sanatorium in Scotland. With a poor prognosis he sailed for New Zealand in 1926. Rest in Fiji and convalescence in New Zealand improved his health, and he gradually returned to his interests in egg collecting and photography.

At various times he was able to travel around the country, particularly the northern half of the North Island, and to many of New Zealand's inshore and offshore islands, joining expeditions or travelling with friends who owned small boats. He also travelled as a paying passenger on the government vessel *Hinemoa* on some of its voyages to service lighthouses and castaway depots.

Between 1943 and 1945, Graham Turbott, the bird curator at Auckland Museum, was absent on war service and Buddle stood in as acting ornithologist. After the war he continued as part-time associate ornithologist until his death in 1951, furthering his interest in oology by cataloguing much of the museum's egg collection. He gave part of his own collection to the museum in 1931; the rest was bequeathed in 1951.

Buddle had kept his eggs in small drawers in a handsome wooden cabinet he may have made himself or commissioned especially for the purpose. Each egg has a coded set-mark written near the hole and Buddle entered these into an exercise book with corresponding collecting details. Like McLean, he also kept field notebooks, and many of these, along with photograph albums and a selection of loose photographs, are in the museum's archives.

Buddle wrote a book, *Bird Secrets*, that was illustrated with his own black and white photographs of New Zealand birds, but he died just before it was published. However, he did see some of his finest bird photographs published in Graham Turbott's book *New Zealand Bird Life*, which was published in 1947. Half a century later, author Steve Braunias illustrated his book *How to Watch a Bird* with Buddle's wonderful photographs and the original captions.

Following his long recuperation, by 1937 Buddle had been well enough to join the New Zealand Total Solar Eclipse Expedition to Canton Island. This atoll was one of the Phoenix Islands, the same scattered archipelago from which John McLean had much earlier obtained booby and other eggs via the Ellis family. Canton, too, had been mined in the 1880s for guano, the accumulated excrement from centuries of breeding seabirds, which was used as a fertiliser around the world. (My own great-great-grandfather, John Simpson, was an agricultural produce merchant in Fife, Scotland, in the 1880s and 1890s and stocked "Peruvian Guano Dissolved & Undissolved".)

Canton lies near the equator, about halfway between Hawai'i and Fiji. A narrow ring of land from fifty to 600 metres wide, it encircles a lagoon of forty square kilometres. The highest point is only seven metres above sea level. It has little rainfall, no freshwater and only sparse vegetation, but it is an ideal site for the breeding colonies of tropical seabirds. Buddle reported only six trees, all coconut palms. In the 1950s the island boasted administrative buildings and a Pan Am hotel; today only a handful of Kiribati families live there.

A total solar eclipse was predicted to be visible from Canton Island on June 8, 1937 and expected to have the longest "maximum totality"—seven minutes four seconds—of any eclipse since AD 699. Both New Zealand and the United States sent eclipse expeditions. The American party was jointly sponsored by the National Geographic Society and the US Navy and characteristically well supplied with men and equipment. It arrived first, from Hawai'i, transported by USS *Avocet*. The New Zealand effort was more of a shoestring affair. The expedition left Auckland for Suva on May 18 aboard RMS *Niagara*. From there the party of eight was taken to Canton Island by HMS *Wellington*, a Royal Navy sloop.

The *Wellington* reached Canton at two in the afternoon on May 26 but the Americans were occupying the only known anchorage. An alternative position had to be found somewhere nearby, but it must have taken time to find, no doubt by depth-sounding, as it was next morning before the New Zealanders landed on Canton with four tons of gear. They established a tent camp alongside the Americans and flew the New Zealand ensign. On the day of the eclipse they were up at forty-thirty a.m. and two naval officers were seconded to help with observations.

It was a tense time. Threatening clouds dispersed at the right moment, and everyone took their stations at their instruments about twenty-five minutes before the outline of the moon first contacted that of the sun. Buddle's task during the event was to draw the outline of the inner corona. In the weird half-light of the early morning eclipse, confused seabirds returned from their recent departure to sea and circled noisily overhead.

Buddle was attached to the expedition to collect specimens for Auckland Museum and about £40 seems to have been paid by the museum for his share of expenses. He was largely released from camp duties during his twelve-day stay at Canton so he could concentrate on collecting. Insects were few and he had trouble finding shells of marine molluscs.

He tried dredging from the expedition's punt, but operating alone he wasn't strong enough to pull the dredge across the sea floor. He lamented the lack of assistance, "all members of the Eclipse party being so busy that I was compelled to work entirely on my own".

The island was overrun with "hundreds of thousands" of Pacific rats: the museum has two of these individuals preserved in alcohol. "The rats were quite tame and at night would sit on the mess table, twenty or thirty at a time, and allow themselves to be stroked and petted ... They finally ate all the labels off our stock of tinned food and when a tin was opened it was a toss-up whether it would turn out to be a tin of tomato soup or a tin of peaches."

Hermit crabs were almost as plentiful as the rats, and the biggest ones hauled shells the size of a closed fist. About camp they were more trouble to Buddle than the rats, and any bird specimen left lying for more than a few minutes would invariably be ruined. Buddle collected fifteen study-skins of birds and between May 27 and June 5 he got sixty-seven eggs, representing eight species of boobies, tropicbirds, frigatebirds and noddies. Three species of boobies were breeding on the island: red-footed (*Sula sula*), masked (*S. dactylatra*) and brown (*S. leucogaster*), some nesting only a few metres from the camp.

Red-footed boobies are the smallest of these three. They nest in low shrubs, lay just one egg, and feed mainly on squids caught at dusk and dawn. The other two boobies are ground-nesters, and while they lay two eggs they almost always succeed in raising just a single chick. They are fish-eaters, the brown booby diving deep for herring-like fish and the masked booby preferring flying fish caught at or near the surface.

Gannets and boobies are members of the order Pelecaniformes, along with the pelicans and shags, or cormorants. These birds have in common the "totipalmate" foot, where three sections of webbing link all four toes. A more common webbed foot is the "palmate", with two webs joining three toes, as in ducks and gulls. In hot weather, gannets and boobies open their beaks, flutter their throats and cool themselves through evaporation from the moist inner lining of the throat.

Both birds are plunge-divers and boobies often work in spectacular feeding groups. Diving from a height, gannets and boobies are beautifully streamlined as they push their extended wings behind them and, dart-like, pierce the surface of the water with their large cone-shaped beaks. As they hit, air sacs inflate under the forward skin to cushion the impact. (Airbags in cars, like many of our inventions, have been anticipated in nature.) The beaks of gannets and boobies have finely serrated cutting edges for seizing fish.

Behind the science of the eclipse expeditions was Great Power rivalry for territory. In April 1937, the month before Buddle arrived, the United Kingdom had annexed the Phoenix Islands to the Gilbert and Ellice Islands Colony, a prelude to its ill-fated Phoenix Islands Settlement Scheme of 1938 to 1963. During the visit of the *Wellington* a sign was erected claiming Canton Island in the name of King George VI. The United States also asserted sovereignty, and appropriate signage was erected by its expedition members.

The New Zealand party began packing up on June 8, as soon as the eclipse was over, and departed Canton that evening on the *Wellington*, bound for Apia in Western Samoa. During the team's twelve-day stay in Apia the photographic plates of the eclipse were developed as soon as possible. The group then boarded the *Maui Pomare* for the homeward journey via Niue, where they had just a few hours ashore. Buddle got back to Lyttelton on July 6 and on July 8 he finished a preliminary report for Auckland Museum. The following year he published an eight-page paper, "Notes on the birds of Canton Island", in *Records of the Auckland Institute and Museum*. The *Wellington* is today moored on the Thames in London as headquarters of the Honourable Company of Master Mariners.

The sovereignty dispute was resolved when Canton Island came under joint UK and US administration in 1939. With the construction of a 1.9-kilometre airstrip, it became an American airbase during the war against Japan, and was bombed once from the air and shelled three times by submarines. After Japan occupied South-east Asia, the island was vital for the defence of Australia and New Zealand: US-built aircraft flew in to equip the air forces of the two countries on the Hawai'i–Canton air route. Canton later became a refuelling point for trans-Pacific civilian flights, until the late 1950s when longer-range aircraft made the Canton stop unnecessary.

During the war 30,000 American military personnel arrived in Canton on rotation. Their presence and the airport's operations had a bad impact on the breeding colonies of seabirds. The eggs and other specimens Buddle collected, and his published notes, came to provide important baseline records for the ornithology of the island.

Auckland Museum has a strong connection to Canton Island through the Buddle eggs, but there is another link, albeit a tenuous one. Buddle's colleague Graham Turbott, later director of Auckland Museum, had a younger brother, Ian, in the British Colonial Service. Ian was serving on Canton Island in 1951 when an attractive American air hostess came down the steps from a Pan Am Boeing Stratocruiser on a brief refuelling stop. Sometime after take-off the plane was forced by engine trouble to return to Canton and the air hostess and her fellow crew members had to stay several days on the island, helping the British and American administrative staff to accommodate the Pan Am passengers. She and Ian were married eight months later. They raised a family and after her husband's long career and honours she became Nancy, Lady Turbott. Research and romance—such are our associations with a once busy tropical South Seas island, now again sunk totally into obscurity.

Song of the huia

In a drawer in the office of the land vertebrates department at Auckland Museum I found, when I started my job, a reel-to-reel magnetic tape. The box was labelled "Mr Archibald Hugh Bogle—Recollections and imitation of Huia call" and dated May 1971.

I didn't have the equipment to play this recording but eventually I took it to a sound technician, who copied it to a cassette tape. The huia, *Heteralocha acutirostris*, was a New Zealand bird that became extinct, probably soon after 1907. Hugh Bogle was recalling birds he had heard around 1903 when he was surveying forest blocks in the Ruahine Ranges. The name "huia" is thought to be an onomatopoeic rendition of one of the bird's shrill whistled calls. Today you can hear historic imitated huia calls—very musical and mournful ones—on the website *New Zealand Birds Online*.

The huia belonged to the family Callaeidae, the New Zealand wattlebirds, so-called because they have a pair of fleshy wattles at the corner of the mouth, one on each side. The wattles were orange in the huia and South Island kōkako. They are blue in the North Island kōkako and red in the North and South Island saddlebacks. Juveniles have the smallest wattles and adult males the largest, so they probably help the birds communicate to each other their age and breeding status.

The huia wasn't just a native bird unique to New Zealand—it and the other New Zealand wattlebirds belong to a family found nowhere else. They are not related to the Australian wattlebirds, which are honeyeaters. Currently, ten families of birds are recognised as being unique to New Zealand: the large number emphasises the country's position as a "biodiversity hotspot". The six families that survive are: kiwi (Apterygidae); kākāpō, kea and kākā (Strigopidae—large parrots); New Zealand wrens (Acanthisittidae—see pages 129–133); New Zealand wattlebirds; stitchbird (Notiomystidae—an unusual songbird); and the whitehead, yellowhead and brown creeper (Mohouidae—small songbirds). The four that are extinct are three families of moa (Emeidae, Megalapterygidae, Dinornithidae) and the adzebills (Aptornithidae), which were large flightless rail-like birds with big pointed beaks.

New Zealand wattlebirds have short rounded wings and fly weakly, but their legs are large and powerful with strong claws, suitable for their largely tree-dwelling existence in forest and scrub. They use their strong bills to probe for arthropods—for example, insects and spiders—and especially like large grubs in decaying wood. They also take berries, and the North Island kōkako eats significant amounts of young leaves.

Of all known birds, the huia had the greatest difference in beak shape between male and female. The bird's latinised generic name *Heteralocha* means "different spouse", a reference to this. The female's bill was much longer, up to 104 millimetres, and therefore more curved than the male's, which didn't exceed about sixty millimetres. This suggests the sexes fed in different ways. Naturalists who watched huia feeding recorded that the males tended to chisel at the outer surfaces of decaying or live wood, whereas the long-billed females were able to probe deeply into soft wood or crevices to retrieve insects unobtainable to the male. The Natural History Museum in London has several huia preserved whole in alcohol. An ornithologist at the museum, Philip Burton, published a study in 1974 in which he described dissecting the head and neck of some of these museum specimens to examine the muscles. He showed that the male had a capacity for forceful "gaping": it could insert its bill into rotting wood and force open its beak to split the wood.

The huia was the largest New Zealand songbird at the time of European settlement. It was about the size of an Australian magpie and weighed around 300 grams. The two species of native ravens were three times heavier, about 950 grams, but these birds had been eliminated by the impact of Polynesian settlement. Today the introduced rook, weighing about 400 grams, is the largest songbird in New Zealand. The biggest native songbird is the huia's relative the North Island kōkako at around 230 grams.

The writings of New Zealand naturalists of the late nineteenth century indicate that the huia frequented deep forest and forest edges, seldom flying but rather moving in leaps and bounds in the canopy or at lower levels. Little is known about its breeding habits. The saucer-shaped nest was large, as expected for so big a bird. It was about 350 millimetres across with a depression for the eggs, lined with fine material. The breeding season was early summer, which is when most birds breed. Two to four eggs were laid, each about forty-five by thirty millimetres, and greyish with brown and purplish markings. Canterbury Museum has the only surviving nest and the Museum of New Zealand Te Papa Tongarewa the only egg.

The huia was unusual in being restricted to the North Island; there is no evidence of its existence elsewhere. Most New Zealand forest birds are present on both North and South Islands. Or, if they are present on only one, they usually have a closely related counterpart on the other. There was no South Island equivalent of the huia.

At the time of European settlement, huia were reliably recorded only in the southern part of the North Island, from the Raukumara Range and Turakina River south to Wellington. The bird had been missed by the naturalists on Cook's voyages, presumably because they spent little time in its restricted range. The first published reference was not until 1835 in Reverend William Yate's *An Account of New Zealand; and of the Formation and Progress of the Church Missionary Society's Mission in the Northern Island*. Yate was in New Zealand from 1828 to 1834, mostly in the Bay of Islands. His book had no official sanction and therefore upset church officials, but it contained a chapter on natural history. Yate mentioned thirty-three kinds of birds by their Māori names, one of the earliest published lists of New Zealand birds, but his information on the huia was second-hand: he never travelled to the southern areas.

In 1837, in *A Synopsis of the Birds of Australia, and the Adjacent Islands*, John Gould in London was first to describe and name the huia to zoological standards. He had only two partial specimens, which lacked feet and much of their wings. They had come to the Zoological Society of London from a sea captain, who had got them from a Māori chief. Gould made the mistake of treating the male and female huia, with their different beaks, as separate species, and produced a coloured plate showing just the two heads. He described the long-billed female first, as *Neomorpha acutirostris*, so by the rules of zoological nomenclature this name got priority over the name of the male, *N. crassirostris*, which came later in the text.

As if Gould's error with the sexes were not enough, a similar name, *Neomorphus*, had already been used in 1827 for a group of ground cuckoos from the New World tropics. In 1851 the German ornithologist Jean Cabanis coined the name *Heteralocha* as a substitute for *Neomorpha*, and since 1865 this name has been in general use. However, *Neomorpha* is not exactly the same as *Neomorphus*, and by the rules laid down in more recent versions of the International Code of Zoological Nomenclature it should not have been replaced.

By the late 1950s New Zealand ornithologists had become aware that *Neomorpha* had priority, and in the interests of stability requested the International Commission of Zoological Nomenclature, based in London, to validate the name *Heteralocha* for the huia. The proposal was internationally notified, and when no objections were received the commission voted 23:1 in favour. In 1958, by

Opinion 514, the commission suppressed *Neomorpha* as a name for the huia.

In 1839, Charles Heaphy, Ernst Dieffenbach and two Māori guides made an excursion into the hills overlooking Wellington's Palliser Bay. Heaphy recorded: "On our way homeward the natives suddenly stopped; they heard in the distance the peculiar cry of the huia. Imitating this, and adding a peculiar croak of their own, which they said was very attractive, our guides soon brought two birds—a male and a female—within shooting distance. We abstained from firing for a moment, admiring the elegant movements of these birds as they leaped from tree to tree, peering inquisitively at us, and gradually coming nearer. We now fired with light charges, and brought each a bird down."

The birds were sent to London and enabled John Gould to produce the large and beautiful hand-coloured lithograph of a pair of huia, published in 1841 for the fourth volume of his lavish series *The Birds of Australia*.

The huia was large and spectacular—emblematic of the distinctiveness of New Zealand birds. It was a focus of great interest for the European settlers, and of guilt as it declined to extinction. The country's first set of pictorial postage stamps, issued from 1898, showed both sexes on the three-penny stamp. A female huia featured on the sixpence from 1933 until new decimal coins were introduced in 1967.

There were two main reasons for the huia's decline after 1840: its lowland forest habitat, including "Forty Mile

Bush" in the northern Wairarapa, was cleared for farmland, and introduced predators, initially rats and later stoats and weasels, spread rapidly to all areas. The bird's nests were probably vulnerable, being sited low to the ground. Rats would have destroyed eggs and chicks; the mustelids would have killed incubating and brooding adults.

Internationally, the huia was one of New Zealand's best known birds, and natural history museums and private collectors sought mounted specimens for display, so collectors with hunting rifles were undoubtedly also a factor in reducing so tame a bird at a time when it was already in decline. Regulations making it illegal to kill huia came into force in February 1892, after special urging by the outgoing governor, Lord Onslow, whose second son—the first New Zealand-born vice-regal child—had been named Huia. Enforcement was another matter. When the Duke of York (later King George V) visited Rotorua in 1902 a Māori guide gave him a huia tail feather. He wore it in his hatband, creating an immediate craze that pushed the price of the feathers to £1 each. Typical of songbirds, each huia had twelve tail feathers.

There may have been a fourth contributing factor in the huia's decline. In 1921 George Mason published a report in *Transactions of the New Zealand Institute* on five ticks he had found attached to a wattle of a huia specimen collected near Masterton in the Wairarapa in 1883. The ticks were of two species present in India, and it was surmised they had travelled to New Zealand with common mynas, *Acridotheres tristis*, which had been introduced from India in 1875. It is possible the ticks transmitted a novel avian disease to huia.

There were no reliable sightings of the bird by competent naturalists after horticulturalist William Walter Smith saw three at Mount Holdsworth in the Tararua Ranges north-west of Masterton on December 28, 1907. The huia became firmly fixed in the nation's psyche as a bird wiped out by the Victorians and Edwardians and the general destructiveness of European settlement.

Walter Lawry Buller, the first notable New Zealand-born ornithologist, is sometimes implicated in the huia's demise. Buller was born at the Bay of Islands in 1838, the son of a pioneering missionary. His profession was the law, and he served as a magistrate in the young colony. However, his passion was birds. One mark of a leading natural historian was the production of large-format books dealing with particular groups of animals and plants and illustrated with coloured plates executed by top artists. Buller aspired to put New Zealand birds (and himself) on the world map by writing just such a book. He followed the usual practice of issuing it in parts, sold by subscription, so money could be raised steadily during the expensive project.

Such a high-quality book could not be printed locally. In London Buller enlisted the Dutch artist Johannes Keulemans to draw the birds, using museum specimens as reference for the fine details. The final plates were lithographed and individually hand-coloured. *A History of the Birds of New Zealand* appeared in 1872–73 to great acclaim. A second edition in two volumes was produced in 1887–88, the plates being drawn anew by Keulemans and printed this time by a more mechanised process, chromolithography.

Buller lived in Wellington, and in 1889 the Auckland Museum curator Thomas Cheeseman was acting as agent to distribute copies to Auckland subscribers. Cheeseman's correspondence with Buller reveals a problem: economic depression in the colony meant some subscribers could not pay. Then there was another problem—100 copies were lost in a shipwreck. On Sunday, June 2, 1889 at about 6.20 a.m. the Union Steamship Company's ship *Maitai*, en route from Wellington to Auckland, stuck a hidden rock near Red Mercury Island off the Coromandel Peninsula. The weather at the time was described in newspaper accounts as "thick", and a violent easterly gale was blowing. The seven passengers and ten crew were saved in boats "with the exception of R.H. Larsen, purser, who, losing his presence of mind, refused to leave the ship, and was drowned; also, the forecabin steward, a man known as George [Archer], fell out of one of the boats and was also lost." Larsen couldn't swim and was said to have clung to one of the davits in terror, refusing to drop into a boat or the sea, as many passengers had to, and went down with the ship. The *Maitai* struck the rock with such force that it filled quickly, the hatches burst, and it sank in ten minutes.

The lifeboats made it to land and everyone was later picked up by the government ship *Hinemoa* and brought to Auckland. Three days later, the resident magistrate, J. Giles, conducted a formal investigation into the sinking. No "negligence or want of judgement" was attached to Captain William Dawson and his crew: after the call to abandon ship they had apparently showed "great coolness, promptitude, and judgement". Everything on the steamer was lost, including Buller's books, worth ten guineas each, a total of £1,050.

Buller had done a great service by summarising in his books all that was known about New Zealand birds from his own observations and from details supplied by his correspondents throughout the country. Added to that, the superb illustrations have been endlessly reproduced. I still remember the coloured images from Buller's *Birds* that appeared on the covers of standard-issue stationery items I used as a schoolboy.

Typically for the times, Buller built up his own collection of bird specimens. He shot birds himself and also bought them from shooters who were supplying birds to museums and private collectors. At various times he sold portions of his collection. His specimens have ended up today mainly in four museums, Te Papa in Wellington, Canterbury Museum in Christchurch, the American Museum of Natural History in New York, and Carnegie Museum in Pittsburgh, Pennnsylvania. Auckland Museum has six, including a huia.

Research by Sandy Bartle and Alan Tennyson at Te Papa has shown that Buller's bird collection was relatively small, around 2,230 specimens collected over fifty years. Considering his costs, and the relatively low sums received, Buller is unlikely to have made much profit on the sale of bird skins and may even have lost money. His collecting activities are unlikely to have contributed significantly to the decline or extinction of any birds. For huia, the total number of specimens in his collection was only twenty-four. Nevertheless, his modest collecting has taken wings in the popular imagination: he is unfairly seen as a rapacious destroyer of birds and a significant contributor to bird extinction. As evil personified, he has even been the subject of contemporary artworks and plays.

Museums were, likewise, not a culprit in the huia's demise. A study of genetic diversity has suggested there were around 34,000 huia at the start of European settlement, yet the number of specimens in all the museums of the world totals about 350 by one estimate. When so many huia were being lost to habitat destruction and predatory mammals, and at a time when there was no ability to transfer the birds to predator-free reserves, it is as well that a few were saved for museum collections. Through them we can experience the sad delight of seeing a real huia, albeit stuffed, in a museum display and we have specimens for study. The genetic study was based on slivers of dried toe-pad tissue taken from seventy-three museum specimens, including three collected by Walter Buller and thirteen from Auckland Museum.

New Zealand museums hold about 120 stuffed huia: mounts and study-skins. Many are poorly documented, with no date or collecting locality. This is because they were not collected by or for museums but for private collectors for whom such details were unimportant. After 100 years in a domed glass case on the mantelpiece in a private home, the faded and damaged birds have been handed on to the nearest large museum. Auckland Museum has fifteen mounted huia and sixteen "study-skins". Only three of the latter seem to be true study-skins prepared for a scientific collection; the rest are damaged mounts that have been "relaxed" by a taxidermist into the shape of a study-skin. Only four of the huia have specific collecting localities and even those are vague—“Forty Mile Bush, Wairarapa” for example, or “Ruahine Range”. None has a date of collection.

There is further background to the huia’s story. Late Holocene fossil bones, dating from the last few thousand years, paint a picture of the diversity and abundance of New Zealand birds before the cataclysm—the arrival of humans. Fossil bones of huia have been found not just in the southern North Island, where the bird was known historically, but elsewhere in the North Island, including at sites in Northland and around Waitomo. This indicates the huia began declining after the arrival of Polynesians 750 to 700 years ago, to the extent that it died out entirely in the north. Other birds were affected in a similarly uneven way, including laughing owls, bush wrens and piopio, all of which at the start of European settlement were much rarer in the North Island than the South. This seems to reflect that the density of Māori population was higher in warmer northern areas.

Huia were among the many kinds of birds Māori hunted for food, and they also prized huia beaks and tail feathers for ornamentation. Predation on nests by the Pacific rat or kiore—an Asian species that Polynesians brought to New Zealand—was probably also a major factor in the demise of the huia over more than half the North Island. The huia would have declined to extinction without the intervention of Europeans, something not widely understood, and joined the long list of bird species wiped out by the ecological disruption that followed Māori settlement.

A publisher once asked me to check for accuracy the draft of a children's book about the huia. Probably unwittingly, the author had lent a slight bias to the historical account. Māori use of the huia was romanticised as noble—feathers taken from killed huia were "prized" and "significant"—whereas European activities were regrettable: there were guns, birds were shot and exhibited, and the taint of monetary value was mentioned. In 2006 I saw an exhibition at Te Papa called "Blood, Earth, Fire". The huia featured in one display case and the label gave this account: "Māori hunted the bird for its tail feathers. ... But bird hunting was strictly controlled. When a bird species became scarce in an area, a rāhui (restriction) on hunting was put in place till numbers recovered. The likely final cause of huia extinction was forest clearance by Pākehā (European) settlers ..."

In 2008 Wellington newspaper *The Dominion Post* published an article repeating the argument that early Māori may have been "New Zealand's first environmentalists". It acknowledged there had been mistakes in overharvesting of moa, but claimed the birds were "already under pressure from natural forces". "Fortunately," the article continued, "the pantheon of atua [deities] brought from the homeland provided relationships to all living things. That led to a respect for nature that, combined with the economic incentive and a growing knowledge base, motivated Māori to develop a regime of sustainable practices for natural resources."

As far as birds go, this does not accord with the biological evidence. Moa, and other New Zealand birds such as the huia, were under no adverse natural pressures. They were not declining or senescent or maladapted, as some have suggested. Instead they were as superbly adapted to their environment as any other surviving species on the planet. They just had few adaptations to counter forest burning and mammalian (including human) predation because they had had no need of them.

The scientific evidence is clear and stark. Some thirty-six species of New Zealand birds died out in the roughly 500 years between AD 1250 to 1300, when Polynesians arrived, and 1769, when the first Europeans landed. No climatic or geological changes are known to have occurred during this period that could account for the loss of these birds. Nearly all the exterminated bird species are represented in the bones archaeologists have found in early Māori middens.

New discoveries in the last thirty-five years have shown the New Zealand pattern of bird decimation was a repeat of what happened across the whole of Polynesia as Polynesian settlement spread. South Pacific islands currently have small bird faunas. It used to be thought that the diversity of species was small because few birds had been able to reach the isolated islands. It is now clear the bird communities are remnants of much larger avifaunas that existed before Polynesian settlement.

The first evidence came from the Hawaiian Islands, where American ornithologists Storrs Olson and Helen James studied Holocene fossil bone deposits in the 1970s and '80s. Their analyses disclosed a remarkable and unsuspected extinct Hawaiian avifauna of several dozen new species. They included geese, many of them flightless, flightless rails, two flightless ibises and numerous honeycreepers, small finch-like birds. The bones indicated a wave of extinctions following the arrival of Polynesians in Hawai'i more than 1,000 years ago.

Similarly, bone deposits discovered by American palaeontologist David Steadman and others since the 1980s have revealed numerous extinct species on other islands across Polynesia, including Tonga, the Cook Islands and French Polynesia, and on Fiji and Vanuatu. It seems that humans often reduced the bird diversity on individual islands to less than half. Some have estimated that prehistoric peoples exterminated 1,000 bird species or more in the south-west Pacific. The numbers are high because some birds, such as rails, had evolved separate species on multiple islands.

Evolved over millions of years in a land without mammals, the huia was naïvely tame. It flew only weakly, needed forest, and probably nested close to the ground, inviting predation at the nest. Both of the peoples who settled New Zealand cleared forest, hunted birds and released rats. The environmental impact of any human society, irrespective of the race, culture and technology, was too much for the huia.

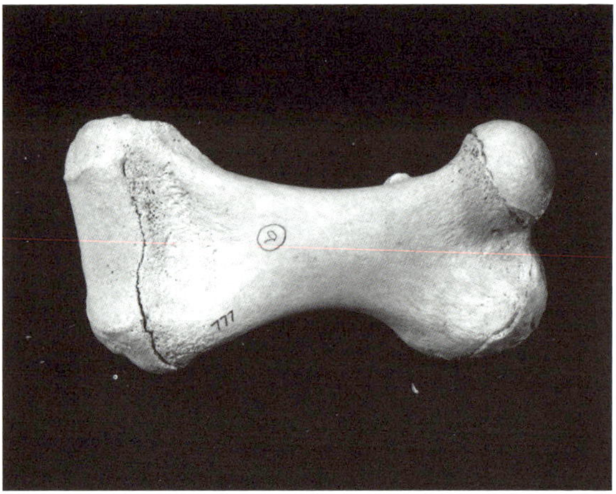

Seals in sand dunes

In a small, acid-free storage box at Auckland Museum there is a partial skeleton of a North Island takahē or notornis, *Porphyrio mantelli*, a large extinct rail. It was found at a sand dune in Northland and now forms part of New Zealand's precious Holocene fossil record of birds. The Holocene (or Post-glacial) Epoch started about 10,000 years ago at the end of the last Ice Age. The takahē bones, and thousands of museum bones like them, are key to understanding the history of New Zealand birds. Each bone gives a geographical data-point in mapping a bird's former distribution.

The rails are a large family of birds found across the globe, and include crakes, moorhens, swamphens and coots. They are ground birds, although most species can fly, and they are chicken-like in their looks and behaviour, although they are not closely related to chickens, which are in the game bird group. Takahē are giant flightless swamphens with massive beaks, and a close relative of the purple swamphen, called pūkeko in New Zealand, which can fly. The bill of swamphens extends on to the forehead as a shield, and both bill and shield are scarlet in the pūkeko and South Island takahē, *P. hochstetteri*.

Between 1849 and 1898, four South Island takahē were seen or taken in the south-west of the South Island. After this the bird was thought to be extinct, until in 1948 a population was rediscovered west of Lake Te Anau by Geoffrey Orbell, a physician. Since then the best efforts of conservation staff, including the hatching of surplus eggs in captivity, have increased numbers of the South Island takahē—the world's largest living rail, weighing two to three kilograms—but it remains endangered. It is vegetarian, eating mainly leaf bases of tussock grasses, grass seeds and fern rhizomes.

The North Island takahē is a taller, slimmer sister species to the South Island one. There is an unconfirmed report of one seen in the Ruahine Ranges in 1894. Otherwise the northern form of takahē has not been recorded historically, one of the many species of birds reduced or eliminated by predation and habitat changes that followed Polynesian settlement of New Zealand.

The takahē and pūkeko seem to represent a "double invasion" of New Zealand by the same ancestral Australian stock. The two takahē descended from an earlier invasion, and had time to evolve a large size and flightlessness. The lineage that continued evolving in Australia colonised New Zealand a second time, and this was so recent that the descendants, the pūkeko, are little different from the Australian purple swamphen. The pūkeko and South Island takahē have diverged too much to be able to interbreed.

Double invasion has been an important mechanism for generating new species among New Zealand birds. Indeed, the North Island and South Island takahē may have arisen by their own separate colonisations, making it a triple invasion.

While the fossil record of New Zealand marine backboned animals—including penguins and whales—is good, the corresponding record for the country's land vertebrates is extremely poor. Consequently, our knowledge of amphibians, terrestrial reptiles and most kinds of birds is nil over most of the epochs since around eighty-five million years ago when, during the break-up of the supercontinent Gondwana, the bit of land that is now New Zealand began to separate from what is now Australia.

It is no coincidence that many of the greatest dinosaur discoveries have been made in North and South America, China and Australia. In these places there are vast, flat, geologically stable areas of sedimentary rocks of the right age that can be searched in arid regions with little vegetation to obscure the fossils. New Zealand, on the other hand, is small, so rocks of any given age, if present at all, may be limited in extent. The land is also geologically turbulent. Earthquakes and volcanoes may speed the exposure of fossils but they also help destroy them. And high rainfall means New Zealand is densely vegetated, with little of the extensive bare ground where fossils are most easily seen.

There are only two major oases in this fossil desert for land vertebrates. One is the Early Miocene site at St Bathans in Central Otago, where palaeontologist Trevor Worthy and his colleagues have recently discovered and described a wealth of fossil bones. The sediments containing these bones were laid down around sixteen to nineteen million years ago in a gigantic shallow freshwater lake ten times the area of Lake Taupō. The climate was subtropical with eucalypts, casuarinas and palms growing on lake shores, along with plants more typical of New Zealand today such as podocarps and southern beech. The site has yielded thousands of bird bones, although unfortunately they are usually isolated ones, rather than associated bones of an individual skeleton, and are often broken. Nevertheless, they have disclosed diverse populations of birds, dominated by ducks but including wading birds, rails, parrots, songbirds and moa. When the site is fully excavated and all the fossils analysed and described, we will have a marvellous insight into the birds of Miocene New Zealand.

Given the general poverty of the fossil record of New Zealand land vertebrates, it is a paradox that there is a fabulous record, one of the best in the world, for the most recent period. Fossilised bird bones from the Holocene Epoch have been recovered the length and breadth of the country—from rock shelters, caves, limestone sinkholes, sand dunes and swamps with alkaline mud. These are sometimes called "subfossils". The bones are nowhere near old enough to have become petrified by geological forces. They are in fact little changed from modern bones, although they are often lighter through having lost their fat (and other biological materials) and a portion of their mineral content.

During a century and a half of scientific investigation, these Holocene fossil bones have painted a dramatic picture of New Zealand's birdlife before the arrival of people. There were nine or ten species of moa; they ranged from turkey-sized to the tallest birds ever known, which could reach three metres when standing upright. Less well known are the large flightless geese, various ducks, Haast's eagle—the biggest ever eagle, large flightless adzebills with savagely pointed beaks, large and small rails including the North Island takahē, an owlet-nightjar, plus New Zealand wrens and large ravens. These species became extinct in the period between when Polynesians arrived and European settlement began.

The fossils also include bones of species, such as huia and New Zealand quail, that died out after European settlement, and of species still living today. The bones show past distributions and relative abundances. Parrot bones—of kakapō, kākā, and red- and yellow-crowned parakeets—are, for example, common at Northland sand-dune sites where no such parrots have lived for a century or more. Comparisons of Holocene bones with those from the preceding Pleistocene Epoch can show how the birds may have changed slightly in size or shape over time. DNA can be isolated from scrapings of fossil bone. This has opened up new avenues of study on the birds' taxonomy and biology.

New Zealand's four main museums are the principal guardians of the Holocene fossil record of birds. Each built up large collections of moa bones from the late 1800s to mid 1900s, when moa attracted much popular and scientific interest, and large quantities of bones were found at famous sites such as Pyramid Valley swamp in North Canterbury. Later, attention shifted to the other, smaller species. Currently, the museums together hold about 77,000 registered lots of Holocene fossil bird bones. This amounts to a salvage operation. Once bones are exposed during the draining and ploughing of swamps, or on the surface of shifting sand dunes, they quickly deteriorate. When they are well cared for in a museum collection, they may last indefinitely.

Coastal sand dunes in New Zealand are fascinating habitats. In a country that is typically damp and green, it is possible to experience in these dunes a blisteringly hot and dry desert-like microclimate. In the 1990s I visited many such sites, often with Fred Brook, who worked for the Department of Conservation. I salvaged bird bones on the surface, looking especially for those of smaller birds poorly represented in Auckland Museum's collection.

New Zealand's coastal sand dunes first formed around 6,500 years ago, when the sea reached its present level and the coastline its present outline. Originally they became forested or covered in scrub. Birds lived and died in and around this vegetation and their bones began to accumulate. After the Polynesians arrived and burnt large tracts of coastal forest it was difficult for this vegetation to re-establish. The habitat changed to what we see today: large areas of mobile, constantly eroding dunes.

When the skeletons of birds are exposed on the dune

surface, the bones move, sliding downhill or getting blown sideways, and become separated from the datable layer in which they originated. The bones of an individual get scattered and the bones of many individuals may get mixed together. Bones may be uncovered, scattered and reburied several times. On the surface, bones already weakened from hundreds or thousands of years in the rain-percolated sand quickly disintegrate through exposure to intense sunlight and the abrasive action of wind-driven sand.

Most bird bones in sand dunes are from natural deaths and not from Māori kitchen middens. This is because natural bones have had several thousand years to accumulate whereas archaeological bones have accumulated during only about 500 years. Even during the archaeological period, only a small proportion of birds in an area were killed and eaten: a much bigger number survived to die naturally.

These fossils from the dunes need to be sorted and identified. My strategy has been to choose a common species, such as the kākā, and tip on to a tray the bones of a whole contemporary specimen from the museum reference collection. Then I go through all the unidentified fossil bones one by one to match them against those of the known bird. Those that match are removed, leaving the rest of the bones to be tested against the next species.

In this way the pile of mystery bones gradually diminishes and a great jigsaw puzzle is solved, although a residue of unidentified bones or fragments usually remains. There is a remedy for any bones that are crumbling or fragile: they can be strengthened by painting them, in a fume cupboard, with a weak solution containing pellets of an acrylic resin,

Paraloid B-72, dissolved in acetone. The acetone quickly evaporates, leaving the bone surface layers hardened by the plastic.

All this effort adds great value to the bones. Once identified—and cleaned and hardened if necessary—they can be assigned registration numbers and catalogued, making them accessible for researchers and an important record of the diversity and distribution of New Zealand's vertebrate animals since the last Ice Age.

In one of his studies in the northern dunes, Fred Brook was trying to determine, by looking at predation on large land snails, when the Pacific rat was introduced to New Zealand. He found predator damage to fossil shells of the large herbivorous flax snail *Placostylus ambagiosus* was at low levels in the sand-dune layers until about 750 years ago, the time of arrival of Polynesian settlers, when there was a marked increase. This supported the view that the Pacific rat arrived with first human settlement, and argued against a radical suggestion that rats were established much earlier, brought by people making impermanent exploratory visits.

In sand dunes in and around North Cape I found bones of seals, both adults and pups. Seal bones had been noted at these sites before but little attention had been paid them and they had been assumed to belong to the New Zealand fur seal, *Arctocephalus forsteri*. This is today the commonest seal around the New Zealand coast. It has breeding colonies in the South Island and southern parts of the North Island, but seals also haul ashore at many places in the northern North Island. Altogether I found bones representing twelve very young seal pups and seventeen adults or large immatures. The pup bones were often in small clusters on the sand, clearly the remains of an individual. Several of these pups were fairly close together in confined areas of the dunes.

The next task was to confirm which seal was involved, since a second species, the New Zealand (or Hooker's) sea lion, *Phocarctos hookeri*, also breeds in New Zealand. This was not straightforward. Like most mammals, including humans, seals have a relatively prolonged "childhood", during which the bones take several years to grow to adult size. This means that there is a big gradation in the size of seal bones, depending on age. (Most birds, by comparison, grow rapidly from chick size to adult size in a matter of weeks so bones of intermediate size are rare.) Adult male seals are also much larger and more muscular than the females, and some male bones—for example, those of the forelimb—develop enlarged ridges and bumps for muscle attachment; these are poorly developed in female bones. All this variation in the bones according to age and sex competes with the subtle differences that may mark out the two species, especially when you are trying to identify damaged or fragmentary fossil bones.

There were few published illustrations of the bones of the two kinds of seals and Auckland Museum's reference collection was limited. We had lots of bones of fur seal pups, from animals that had been found dead on local beaches, but few samples of the adult bones of both species. I had to take my fossil bones to Wellington to use the reference material that was in the care of Anton van Helden, the collection manager for marine mammals at the national museum. Neither museum had bones of sea lion pups, but with the other material I was able to work out the identity of most of my fossils. The results were exciting because all the Holocene seal bones from the North Cape dune sites that I was able to identify belonged to the New Zealand sea lion and none to the fur seal.

The smallest North Cape fossil seal bones would have belonged to young pups or perhaps, in some cases, full-term foetuses. In the late stages of pregnancy New Zealand sea lions gather at rookeries, and newborns remain at these rookeries for their first few months, so they must have been breeding at North Cape. The survival of so many pup bones relative to adult ones, and their close proximity to each other, suggested there had been a substantial breeding colony. New Zealand sea lions breed on or near flat sandy beaches, unlike fur seals, which prefer rocky coasts. Today, New Zealand sea lions breed mainly in the subantarctic Auckland Islands but they have hauling-out grounds around Stewart Island and the southern end of the South Island.

The fossil sea lion bones at North Cape, and similar discoveries at other places, give evidence of a time in the recent past when sea lions bred widely at suitable beaches all around the New Zealand coast. Hunting by Polynesian settlers is the most likely cause of their disappearance. And there has been an interesting twist. Sea lion bones at Auckland Museum, and in other museum and university collections, were recently sampled to recover DNA. The study gave an unexpected result: the sea lions exterminated from mainland New Zealand were genetically distinct, and perhaps a separate species, from those that survive today on the Aucklands.

Ship rats of Big South Cape Island

A dead rat in a jar of preserving fluid is not perhaps the most charming specimen in a museum's natural history collection. Auckland Museum has two particular specimens in alcohol of the kiore or Pacific rat, *Rattus exulans*. They were collected on Taranga (Hen) Island off Northland's east coast by Dick Sibson in December 1948.

"Sib" was a classics master at King's College, a boys' secondary school in South Auckland, and he started the King's College Bird Club to encourage boys into birdwatching. Members got to go on trips. There is another kiore in the museum's collection that Sib obtained on Lady Alice Island in the Hen and Chickens group in December 1962. It was fortunate Sib's camping parties took the trouble to collect these specimens and save them, with their collecting details, for the museum: they have new importance because the rats were eradicated from Lady Alice Island between 1994 and 1997 and from Taranga Island in 2011.

Rodents are one of the great evolutionary successes of the mammal lineage. Some forty percent of the 4,000 living species of mammals are rodents and they inhabit all continents except Antarctica. The order includes rats, mice, squirrels, guinea pigs, porcupines and beavers. The largest is the capybara of South America.

Rodents show great adaptability to different habitats, and diversity in living habits. Different kinds of rodents, to varying degrees, are able to climb trees, live on the ground, burrow beneath the ground, and/or live in and out of fresh water. A characteristic of them all is a pair of sharp incisors on both the upper and lower jaws, which grow continuously to replace what is worn away by gnawing. Indeed, rodents are named for their gnawing: the Latin *rodere* means to gnaw or eat away.

Rats, along with mice and gerbils, belong to the family Muridae. There are about a thousand species of murids, making it the largest mammalian family. Rats are the only ground-dwelling placental mammals that spread to Australia of their own accord, dispersing from the north, but they reached New Zealand just recently, and only with human assistance. The rat specimens in Auckland Museum's collection are a pointer to one of the big stories of New Zealand biogeography—the arrival of land mammals in a previously mammal-free country.

The chunk of land that is now New Zealand once sat at the edge of the southern hemisphere supercontinent called Gondwana after a region in India where important rocks relating to the supercontinent were found. As the Tasman Sea between New Zealand and Australia steadily widened, New

Zealand began to break away to become the world's most isolated land mass of its size. The latest research indicates this separation began in the south about eighty-five million years ago, but was not complete in the north until fifty-five to sixty million years ago.

The extreme isolation helped shape the distinctiveness of New Zealand's endemic animals. Many potential competitors that evolved elsewhere (like modern rats) were barred from dispersing to New Zealand by the water barrier, and old lineages of animals that died out elsewhere were able to persist in New Zealand without competition.

Scientists long assumed that the separation of New Zealand from Gondwana was too early to have allowed the spread of land mammals, which are represented today in the native fauna only by seals (whose ancestors swam to New Zealand) and bats (whose ancestors flew there). That idea has had to be revised a little following fossil discoveries at St Bathans in the South Island. Among thousands of small fossil bones and fragments sixteen to nineteen million years old, the St Bathans team found two jaw fragments from mouse-sized land mammals. This fragmentary evidence questions the notion that New Zealand never had land mammals.

But for whatever reason these animals did not survive to the Holocene—the fossil record for this period shows no terrestrial mammals except seals and bats. In fact the St Bathans land mammals can't have survived much beyond sixteen million years ago because some New Zealand birds, such as kiwi, takahē, weka and kākāpō, evolved large size and/or flightlessness, implying they lived a very long time without predatory mammals.

The dying out of dinosaurs and pterosaurs, large flying reptiles, sixty-five million years ago had a huge effect. Birds, the descendants of one line of dinosaurs, and mammals, the descendants of a very different reptilian line, became free to diversify and become the dominant vertebrate animals on land. For tens of millions of years birds have dominated the skies during daytime, with bats and some nocturnal birds taking over at night. Birds are driven by natural selection to be lightly built and able to fly, in part so they can escape from predatory ground-dwelling mammals.

New Zealand represents a natural experiment on a grand scale. If the mammals of the world had died out along with the dinosaurs, and if birds had come to dominate land habitats unchallenged, how would birds have evolved? The development of New Zealand birds in the absence of mammals has provided some answers. Released from the pressure of mammalian predation, some birds become large, flightless and long-lived, and raise small families slowly.

The return of mammals to New Zealand was a calamity for the birds, which had become naïve and had no adaptations to counter being killed by mammals. It started in about AD 1280 with the arrival of *Homo sapiens*, the most formidable and destructive mammalian pest of all. The first were the Polynesians, who came on large wooden sailing canoes at the end of a five- to six-thousand-year-long territorial expansion that took them from their home in Southeast Asia, across the islands of the south-west Pacific to settlement in Hawai‘i and

New Zealand. Within a century at most, Māori settlers had burned and cleared large tracts of forests, including about half those in the South Island. They brought domestic dogs and kiore, a rat at home in Asian paddy fields but transported accidentally (or perhaps in some cases deliberately) to most islands of the south-west Pacific. Burning, hunting, dogs and rats exterminated many birds.

Next came the Europeans. The first to set foot in New Zealand were members of James Cook's first voyage of discovery in 1769. On January 17, 1770, when the barque HMS *Endeavour* was anchored at Queen Charlotte Sound in the north-east of the South Island, the naturalist Joseph Banks wrote in his journal of the dawn chorus of the bellbirds: "This Morn I was awaked by the singing of the Birds ashore from whence we are distant not a quarter of a mile, the numbers of them were certainly very great who seem'd to strain their throats with emulation perhaps; their voices were certainly the most melodious Wild Musick I have ever heard, almost imitating small Bells but with the most tuneable silver sound imaginable ..."

Unfortunately, like all ships of the period, *Endeavour* was infested with rats, probably of the Norway, or brown, variety, *Rattus norvegicus*. Even while Banks was enjoying the birdsong the rats were no doubt spreading ashore, the start of a sequence of events that would further diminish New Zealand's "wild musick".

After two more expeditions by Cook, ships of whalers and sealers from Europe and North America came in ever increasing frequency, ensuring the establishment and spread not only of Norway rats but of house mice and domestic cats. Ship, or black, rats, *Rattus rattus*, probably came a little later, in the 1860s, and in the late 1880s three mustelids—stoat, weasel and feral ferret—and the hedgehog were deliberately introduced from Europe. Europeans also brought herbivorous mammals, notably brush-tailed possums, pigs and various deer, many of which spread and began to harm the native forest. European technology allowed the clearing of forest and the draining of swamps and wetlands with increased efficiency. Following European settlement another sixteen species of New Zealand birds were eliminated, the last, the bush wren, in 1972.

The kiore of the Hen and Chickens Islands were descendants of the original rats from Southeast Asia, and the Auckland Museum specimens from 1948 and 1962 document animal populations that no longer exist. Today New Zealand ecologists and field staff are world leaders in eradicating rats and other invasive mammals from islands. As techniques have improved, it has been possible to wipe out rats on larger and larger islands, so increasing the number of places suitable as reserves for the country's rare and endangered animals and plants. To date the largest New Zealand island made pest-free is subantarctic Campbell Island: its 11,300 hectares were cleared of Norway rats in 2001.

Collections of local land mammals in New Zealand museums are dominated, of necessity, by introduced species. My predecessor at Auckland Museum, Sylvia Reed, saw the value of collecting local mammals and began to increase the holdings. When she received a batch of dead kiore from Little Barrier Island, she had them prepared by a taxidermist as a series of study-skins and they are a permanent and valuable record of the population now that island has been made pest-free.

I, too, was anxious not to neglect the introduced mammals and had many opportunities to expand the museum's holdings. One easy addition came from Auckland University: a collection of skulls of wallabies collected on Kawau Island around 1970 by a post-graduate student. Two mammalogists, John Innes and Carolyn King, offered scores of rat skulls from their research in the central North Island in the early 1980s. In 1996 we asked Department of Conservation offices around the country for bones, and Shane Cross of Hokitika, a keen deerstalker, provided several skeletons of red deer, thar and chamois—species missing from our collection. He shot the animals and left them in the bush to rot, later recovering the bones and freighting them north for us to clean.

The growing mammal collection has often been put to good use. In 2013 a biologist, George Lyras, from Athens' National and Kapodistrian University arranged to examine some of our mammal skulls during a visit to New Zealand. I apologised for the small size of our collection but he was complimentary. It seems many of the northern hemisphere museums had neglected their small number of introduced mammals in favour of native ones, to the detriment of research. Animals can change in a new environment and George was studying how mammals on islands have evolved. In the Mediterranean, populations of dwarf elephants developed on some islands in prehistoric times. The way mammals have changed in size, albeit minutely, following their arrival on mainland New Zealand, and also on our small islands from the mainland, can throw new light on the processes of evolution.

In the 1990s I set up a special cupboard in the storeroom of Auckland Museum's land vertebrates collection, and moved most of the stuffed specimens of extinct birds into it. Here it would be easier to control access to these precious objects, and prevent casual handling, than if they were left scattered among the main groups of specimens. In 2001 I heard that wildlife authorities in Hawai'i had declared extinct a bird called the kaua'i 'ō'ō, whose numbers had dropped below ten in the 1980s. I knew we had a specimen that had been collected on the island of Kaua'i in January 1893 by George Munro, the Hawaiian ornithologist who had grown up in Clevedon near Auckland. I had the sad task of moving this study-skin from the main collection to the Extinct Birds Cabinet.

Extinction is the fate of every species. Some die out and their evolutionary line reaches a dead end. This happened to nearly all the dinosaurs, which abruptly disappeared from the fossil record at the end of the Cretaceous Period. But many extinctions are not as grim, because the species die out as their descendants evolve, generation by generation, into one or more slightly different species. This is extinction without cataclysm: the lineage continues unbroken even though in the fullness of time an antecedent species has disappeared.

Extinction may be natural, but a new development in Earth's history is that the rate of extinction has been accelerated to unnatural levels wherever humans have spread. Auckland Museum's Extinct Birds Cabinet is particularly sad because all the birds in it have been hastened to extinction by human activity. On one of the trays, resting on a padded acid-free lining especially installed by a museum conservator, there are two very small drab birds. Their modesty belies their importance for they are bush wrens, *Xenicus longipes*, belonging to the New Zealand wren family. One was obtained from a Mr W.H. Combes in 1881 and is said to be from north-west Nelson. The other, acquired in 1928, has no locality recorded. Bush wrens were close relatives of the rock wren, *X. gilviventris*, which survives as a threatened species in alpine areas of the South Island. Another more familiar relative is the rifleman, *Acanthisitta chloris*—New Zealand's smallest bird—which can be seen in native bush, and sometimes in scrub and exotic pine plantations adjoining native forest, in many parts of the country.

Lyall's wren, *Traversia lyalli*, was discovered in 1892 on tiny Stephens Island in Cook Strait. By 1895 it was extinct—possibly the shortest interval between discovery and extinction of any animal. In 1894 and 1895 one of the lighthouse keepers, David Lyall, retrieved from the keepers' domestic cats about seventeen specimens of the wren, a dozen of which are still in natural history museums in several countries. Since then, Holocene fossil bones of this species have been found at several sites on both the North and South

Islands, showing that the population on Stephens Island was the vestige of a once widespread species almost eliminated by Māori impact.

Three further species of New Zealand wrens have been described from fossil bones recovered from Late Pleistocene and Holocene caves and sand-dune deposits. These died out following the arrival of Polynesians. There was the North Island stoutlegged wren, *Pachyplichas jagmi*, and the slightly larger South Island stoutlegged wren, *P. yaldwyni*. The long-billed wren, *Dendroscansor decurvirostris*, is the only New Zealand wren with a significantly curved beak, which curves downwards, and so far its bones have been found only in the South Island. These three species and Lyall's wren were all apparently flightless, the world's only flightless perching birds, or "passerines". Lyall's wren was the world's smallest flightless bird.

The New Zealand wrens have a zoological importance out of all proportion to their restricted world distribution, limited number of species, faint calls, small size and drabness. The latest DNA studies show they are the most "primitive" living passerines in the sense that their ancestors diverged earliest from the other passerine lineages. They are in other words the sister group to all the other living passerines, including all the world's songbirds. This supports the idea that passerines originated in Gondwana. New Zealand wrens are survivors of a less advanced type of passerine, which died out in many other parts of the world with the evolution of songbirds.

The bush wren weighed about the same as a European goldfinch and was considerably smaller than a house sparrow. It had particularly long legs, hence its Latinised scientific name "longipes". It was a weak flier and often fed on the ground. It had a habit of bobbing its head and body, especially on landing. Nests were hidden in cavities among tree roots, fallen logs or clumps of ferns. Both parents incubated the eggs and fed the nestlings. Bush wrens had no song but gave faint rasping sounds or loud cheeps of alarm.

The North Island subspecies of bush wren was so rare in the time since historical records began that there are only three specimens in museums, two collected in the Rimutaka Ranges near Wellington about 1850 and one from Taupō in the central North Island. Two were seen near Wellington in 1918. The last reasonably sound sightings were near Lake Waikaremoana, east of Taupō, in 1949 and 1955. The South Island subspecies was widespread at the time of early European settlement, especially in mountain forest. The last probable sightings were in Southland in 1947 and the Nelson Lakes National Park in 1968.

The Stewart Island subspecies was studied in the 1930s on islands off the southwest coast of Stewart Island. There were possible sightings of bush wrens on the Stewart Island mainland in 1950 and 1951, but by the 1960s it had become confined to just three outliers: Big South Cape Island (940 hectares) and two nearby islets, Solomon (twenty-six hectares) and Pukeweka (four hectares). These three are "mutton-bird" islands, places where sooty shearwaters nest in burrows. Local Māori families traditionally visit in autumn to collect the large downy shearwater nestlings, the mutton-birds, for food. Big South Cape Island was covered in low forest and largely unmodified except for areas occupied by the huts of the twenty-five or so families who spent up to three months on the island each year.

In 1955 a scientific party trapped a ship rat on Big South Cape. This must have been an isolated individual because the mutton-birders saw no damage to the soap, candles, bags of flour and other items they were accustomed to leaving exposed in their huts. More rats must have reached the island. In autumn of 1963 a mutton-birder noticed some potatoes in his hut had been chewed by rats. Next season damage by rats was more widespread and in March 1964 news of the problem reached the Wildlife Service in Wellington. The officers who arrived to investigate in April found the birdlife and large insects in the north of the island greatly depleted. The service placed bait stations containing rat poison near some of the huts on Big South Cape and all over Solomon Island. The poison may have delayed the spread of the rats but soon they were everywhere. Robins and fernbirds quickly died out and other birds declined.

In August and September the Wildlife officers transferred some birds of the rarest species from Big South Cape to nearby rat-free islands. Six bush wrens were caught for transfer to tiny Kaimohu Island but it was difficult to keep them in captivity and three died. In a final push, three more were caught and a rendezvous was arranged for the next morning with a Royal New Zealand Navy fisheries protection launch. The seas were too rough for a dinghy to land on Kaimohu. Instead, the launch attempted to nose up in the swell towards a rocky shelf. Brian Bell, leader of the Wildlife Service party, bravely jumped ashore. Boxes containing the wrens were landed by running them from the boat along a taut line and Bell set the wrens free in the nearby scrub. To get back he attached the line to his life jacket, jumped in the cold swell and fought to stay afloat until he was hauled aboard.

Bird translocations were a new idea then, and the work was pioneering and experimental. Wildlife officers were back on Big South Cape in February 1965 but could find no wrens for transfer. Two were seen on Kaimohu in 1972 but an inspection in 1977 and subsequent visits revealed none. It is clear now that these had been the last of the bush wrens and their absence had signalled the extinction of the species.

The loss of this wren is achingly sad at several levels. It is the most recent loss of a species of New Zealand bird, and the wren was imperilled just a little too early to be saved by translocation methods now perfected. Its demise, not realised at first, attracted no public attention, and it was the loss of a member of a family important internationally to the understanding of bird evolution.

The rat plague on Big South Cape Island did even more damage: it eliminated a population of snipes now regarded as the last survivors of the South Island snipe, *Coenocorypha iredalei*. Two snipes were caught on Big South Cape in August 1964 but died as bad weather was preventing their transfer to a rat-free island. Also exterminated by the rat invasion was the greater short-tailed bat, *Mystacina robusta*.

However, there was a good-news story. In 1964 thirty-six South Island saddlebacks, *Philesturnus carunculatus*, were caught on Big South Cape and transferred to Kaimohu and Big Islands. More were rescued in 1965. Numbers on the new islands increased and in the last fifty years this species has been translocated to, and established on, many southern islands. The North Island saddleback, *Philesturnus rufusater*, was saved in a similar way. The saving of these birds was a world first—the first time a bird in danger of extinction was restored to viability by direct intervention.

Ship rats probably reached Big South Cape Island along the mooring lines of fishing boats used to transfer mutton-birders to the island. As the island had been rodent-free, the rats had no competitors. There were also few other predators on the island, just weka (large rails) and moreporks (owls).

The explosion in numbers of ship rats on the island was a phenomenon called an irruption. As the first rats multiplied and spread, the island would have been a paradise for them, with its abundant food and few privations. Ship rats are generalist omnivores and adapt to whatever food is available. In the New Zealand forest they will eat fruits and seeds, shoots of plants, insects, snails and lizards. Unlike Norway rats, they are competent climbers and can reach birds' nests in tree cavities or towards the ends of outer branches. They eat nestlings, and birds' eggs up to at least sixty millimetres long. Often they will succeed in catching and killing an adult small bird on the nest.

Ship rats can breed at three to four months old and produce five to eight young in a litter, with multiple litters per year. Under the favourable conditions on the island the initial rat population would have begun to increase exponentially. This is a frightening form of growth that starts slowly, gathers momentum, then turns a corner and burgeons unbelievably. A chessboard has sixty-four squares. If you place one grain of rice on the first square, two on the second, four on the third, eight on the fourth, sixteen on the fifth, and so on, this demonstrates a form of exponential growth. It's easy at first and everything seems manageable, but it soon becomes a nightmare. On the twenty-first square you need to place more than a million grains of rice. You need more than a million million grains for square forty-one, and there isn't enough rice on the planet to complete the last squares.

Very soon the whole of Big South Cape Island was swarming with rats. All horizontal surfaces in the huts of the mutton-birders were fouled with rat droppings, and rats were nesting among their bedding and inside their mattresses. Many kinds of birds were fast declining as they and their breeding attempts succumbed to rat predation. The largest and most accessible kinds of insects were also hard to find. Fallen berries had all been eaten, certain kinds of highly palatable plants were defoliated, and any edible seedlings on the forest floor were chewed to ground level. The island was becoming a hellhole.

Nature's feedback mechanisms now increased in proportion to the problem. As hunger prevailed, the condition of the rats would have deteriorated. They lived shorter lives, and reproduced less prolifically. The fewer litters of fewer offspring contained many sickly young. At peak density the rats would have been in constant contact, with high levels of aggression and fighting. Stress hormones were at unhealthy levels and there must have been cannibalism. High rat numbers would have allowed any infectious diseases and transmissible parasites to quickly run rampant through the whole population. In the huts, the desperate rats stripped wallpaper off the walls to eat the flour paste. Rat numbers then crashed cataclysmically.

But some rats survived. As the vegetation recovered and the prey animals that had not been eliminated multiplied again, rat numbers would have increased. This time the rat population remained in equilibrium with the island's degraded resources. Finally, a special poisoning campaign—using the techniques attempted on the island in 1966 but now improved—saw rats eliminated in 2006.

The events on Big South Cape Island show in microcosm what happened several times across the whole of New Zealand as native birds faced the arrival and spread of new predatory mammals. But the story is even more instructive: the events reflect what is happening over the whole planet. People are the ship rats. Big South Cape Island is the Earth, a piece of territory of finite size, able to support only so many people, with a limit on human numbers and economic growth. We have occupied all the world's habitable regions. Our population has reached about 7.4 billion. Our numbers are now in the dreaded phase of exponential growth, rocketing uncontrolled and uncontrollably to an uncertain future. Every second, an average of 1.8 people die—but another 4.3 babies are born. Some predictions are that the human population will soon level out but not before rising by more than fifty percent. Whatever happens, there are clearly too many of us and our activities are causing too much damage to the environment on which we depend. Some put their faith in technology to solve all the problems. Realists doubt that human ingenuity can cope when it would have to produce the required innovations at an accelerating rate—indefinitely.

Pity the poor ship rats of Big South Cape Island in 1965. Their tiny brains gave them no idea of what they were doing, nor any inkling of what lay ahead. Blindly they marched towards the abyss. But we are advanced primates with strange spherical skulls enclosing the enlarged cerebral hemispheres that natural selection shaped to give us startling cognitive abilities. We can see the problem. Alas, this evolutionary "gift" is cancelled out by fatal flaws in our disposition that we share with the common chimpanzee, our cousin on our branch of the anthropoid apes. We are tribal, sometimes considerate and self-sacrificing towards our own family, ethnic, religious or national groups but often hostile to others. We, and especially our testosterone-fuelled young men, are frequently aggressive and violent. We kill our own kind—periodically en masse in a sort of industrialised carnage. Our tribal self-interest and greed for short-term gains count against our ever agreeing globally on a rescue plan.

Countries want the impossible—economic growth without end—and our main politico-economic structures promote few alternatives. Yet endless economic, material and population growth is as damaging and pointless as was the proliferation of rats on Big South Cape Island.

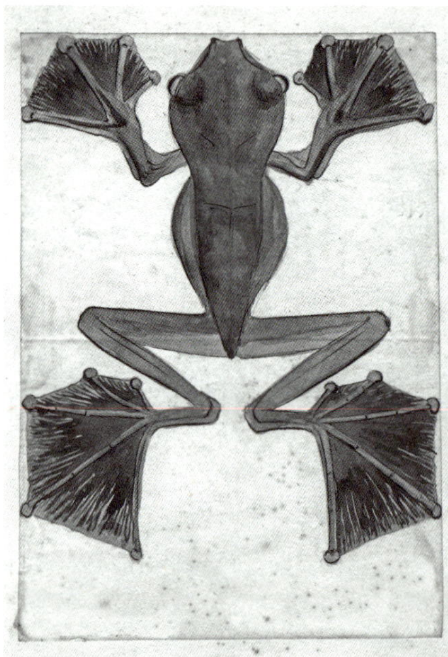

Charles McCann's giant flying frogs

During a tea break at a scientific conference in 1980 I was introduced to an elderly white-haired man who was resting on a park bench in the university grounds. This was Charles McCann, a big name in New Zealand herpetology, the study of amphibians and reptiles. The conference was a three-day symposium, held at Victoria University of Wellington. More than sixty people attended; it was the first large conference devoted to New Zealand "herps", then a rather neglected and poorly known element in our fauna. A recent graduate, I was lucky to attend such a meeting. I was living near Wellington and had time on my hands before moving to Brisbane to take up a research fellowship. Charles McCann attended parts of the symposium but later that year he died, just five days short of his eighty-first birthday.

Yule Mervyn Charles McCann, born in 1899, joined the staff of what was then the Dominion Museum in Wellington in 1948, and served as a zoologist there until he retired in 1964. One of his first big projects was a study of the lizards—geckos and skinks—of New Zealand. He examined preserved specimens in the country's museums. In 1955 he published *Dominion Museum Bulletin No. 17*, a 127-page monograph entitled "The lizards of New Zealand—Gekkonidae and Scincidae". In this document he listed the species he considered valid, described and named about ten new species, and summarised what was known about the distribution and habits of each kind of lizard.

Details of the 1980 symposium were published in a volume dedicated to McCann and his biological career. Delegates didn't learn until afterwards, and too late to help, that this leading light in our field had been struggling financially in retirement, forced to work as a petrol pump attendant in his late seventies.

Twenty-nine years after the Wellington meeting I attended another herpetological conference, this time at Massey University in Auckland. In a presentation I pointed out that the herpetology collections at New Zealand natural history museums were a resource for research but herpetology had been long neglected and the total holdings of about 11,500 specimens were miniscule compared to public collections in comparable countries such as Australia. I made a plea for researchers to salvage lizards that died accidentally, properly preserve and fully document them, and lodge them in our public collections.

At morning tea I met David Chapple, a researcher at Victoria University, who told me the university's school of biological sciences had in storage Charles McCann's research collection of Indian amphibians and reptiles, which was more appropriate to a public natural history museum that would make it more readily accessible. McCann had worked in India before he emigrated to New Zealand: the Bombay

Natural History Society had later established a fieldwork fund in his name.

In due course some fifty-five jars of specimens were transferred to Auckland. These proved to contain about 540 individual animals, large and small, a wide representation of Indian herpetofauna, from caecilians—primitive worm-like amphibians—to snakes. It included many geckos and skinks, species relevant to New Zealand's own fauna. Today the collection complements Auckland Museum's large holdings of Australian and Pacific herps, which are sometimes used to help identify specimens intercepted at the New Zealand border by biosecurity officers.

For museum collections, amphibians and reptiles are typically preserved whole in fluid. As soon as possible after the animal dies (or is defrosted, if the dead animal has been frozen), it is "fixed" using a weak aqueous solution of formalin, which is itself a forty percent solution of formaldehyde gas in water. The belly is slit to admit the fixative, or the fixative injected into the abdomen and into large areas of tissue such as the base of the legs and tail. The animal is then submerged in the fixative for a few hours, days or weeks, depending on size, until it becomes somewhat firm and stiff. The formalin acts quickly to kill bacteria throughout the animal, and therefore stop decay. It also stops the animal's own internal chemical reactions from destroying the corpse.

Because the animal will be stiff, it is important to arrange it in a desired posture beforehand. The straighter the body the easier will it be to measure its length: the standard measurement is from "snout to vent". If the fingers and toes are straightened and arranged neatly, it will be easier to later examine the pattern of scales.

After fixation, the animal is soaked or washed briefly in plain water and transferred to ethanol, usually seventy percent, for long-term storage as a "spirit specimen". Ethanol, otherwise known as ethyl alcohol, is of course the alcohol we humans drink from time to time. Despite being stored in dark cupboards, spirit specimens tend to lose their colour and the McCann specimens are now mostly various shades of brown.

Amphibians and reptiles preserved in spirit will last for centuries if they are cared for properly. In the bird collection of the Natural History Museum in Hertfordshire, England there is a a spirit specimen of the South Island saddleback, a blackbird-sized bird, that was most likely collected on James Cook's first voyage to New Zealand more than 240 years ago.

When herpetologists visit museums they are given some bench space close to the collection. The specimens of interest are retrieved from storage. The researcher may then remove each animal from its jar and place it on a tray, where it is measured and surface details examined, if necessary under a stereoscopic dissecting microscope. Spirit specimens may be X-rayed to reveal the bones inside, or dissected to examine the internal anatomy. Small tissue samples may be detached for DNA analysis and the specimens stand ready to be submitted to whatever techniques of analysis may be devised in future. Each specimen must have a registration number that links to the details—critical for any study—of where and when it was collected.

McCann was particularly interested in reproduction. He preserved many of his specimens dissected out to expose the internal gonads: testes in the male and ovaries and oviducts in the female. Like birds (but unlike mammals), amphibians and reptiles possess a cloaca, a common opening for products of the intestinal, urinary and reproductive tracts. The main sewage system in ancient Rome was the Cloaca Maxima, and the early anatomists, with their classical knowledge, gave the name to the zoological orifice that performed a vaguely similar function. In reptiles the cloaca opens to the exterior near the base of the tail, at a transverse slit called the vent.

Instead of a single penis, male snakes and lizards have paired hemipenes, or "half-penises". These are normally retracted into the base of the tail, but during sexual excitement each can pop out and protrude from the vent, rather like when a finger of a rubber glove is reversed and you blow air into the glove to make the finger "evert".

McCann made a special study of the hemipenes of Indian snakes and lizards and published an important paper on this subject just before leaving India. "The hemipenis in reptiles" appeared in the *Journal of the Bombay Natural History Society* in 1946. "I adopted a simple process of everting the organ when in the fresh condition, and of fixing it in the everted condition in preservation," McCann wrote. Some of these specimens are in the Auckland Museum collection. The everted snake hemipenes are rather bizarre and alarming to behold, with their various fissures, furrows, spines and excrescences.

McCann drew and described a selection of hemipenes for his paper: differences among them can prove useful in separating closely related species. Many snake hemipenes look spiky. As McCann pointed out, their dilated heads "establish a firm copulatory hold within the female, thus dispensing with the need for an external hold during coitus". You can sympathise with a snake's attempt at holding their mate when they have no arms and legs. McCann found it difficult to observe whether one hemipenis was used for copulation, or both at once. Today the consensus is that males use only one at a time.

Charles McCann was born in 1899 at Castlerock in India's princely state of Mysore—then the international frontier town between British India and the Portuguese enclave of Goa. His father had been in the Indian Civil Service and his grandfather had headed the Indian Lighthouse Service. He passed his childhood in country districts close to the local people, so he became well grounded in India's nature and culture. After attending a Jesuit high school in Bombay he was employed from 1916 to 1920 at St. Xavier's College, Bombay, first as a laboratory assistant and later as a fellow. He was able to take, and later instruct, tertiary classes in biology, and also undertook research with Father Ethelbert Blatter, a Swiss Jesuit and pioneering botanist who founded a herbarium that is still one of the most important in India. My guess is that McCann's son Carlyle Ethelbert, born in 1928, was named after Blatter.

In 1921 McCann joined the Bombay Natural History Society as a field collector; he would later become assistant curator. The society, founded in 1883, is a venerable scientific organisation. McCann was a prolific contributor of papers and short notes to the society's journal and co-edited it from 1932 to 1947. Given his roots in India and his established reputation as an Indian biologist, he was no doubt distressed by the events leading up to 1947, when India gained independence and future prospects for Europeans on the subcontinent became uncertain. He decided to emigrate to New Zealand with his wife Eleanor, his three children Carl, Trevor and Marie, and must have brought with him his precious personal collection of Indian snakes, lizards and amphibians.The family left India on the *Chyebassa*, a cargo ship carrying tea and a small group of passengers. In Sydney they transferred to the *Wahine* and reached Wellington on October 7, 1947.

I worked on the McCann collection intermittently for more than a year with Jason Froggatt, a collection manager. Most specimens had paper labels tied to them; some smaller specimens housed inside corked glass vials had loose labels enclosed. Details on the labels had been written in pencil. Jason and I opened the jars in succession and sorted carefully through the specimens, removing and air-drying the labels and deciphering the details. In the case of place names and species identifications we had to find the current equivalents. This was quite a task. A label might, for example, have something that looked like "Andur", a place in Tamil Nadu, but when we looked at the handwriting on more labels it became clear the name was "Andheri", a suburb of Bombay.

In this way the project slowly proceeded.

We attached to each animal a unique Auckland Museum numbered tag, recorded the details on individual catalogue cards, and entered them into the museum's electronic collection database. The animals, grouped by species, went into new jars and we shelved them in taxonomic order with the rest of the museum's herpetology collection. We wrote the new registration numbers on the respective McCann labels as a cross-reference and kept all original labels for future reference.

We also wrote an account of the collection for *Records of the Auckland Museum*. To put the collection in context I needed to read all twenty or so articles on Indian herpetology that McCann had written for *Journal of the Bombay Natural History Society*. Some of the journal's articles were available in scanned electronic form on the internet but not for the decades I was interested in. After looking in vain for the journal in Auckland Museum library's electronic catalogue, I asked a librarian for help. She told me the museum had some but they were in a basement, wrapped in plastic, with cubic metres of other material, and the volume numbers were not listed. They would not be breaking them out on the off chance the relevant issues were present. However, the journal was, she said, held at the libraries of both Te Papa Tongarewa and the National Institute of Water and Atmospheric Research (NIWA) in Wellington.

Te Papa library's online catalogue showed almost all the volumes I needed were there, but when I emailed to set a visiting time I was told that in fact none was present. I had more luck at NIWA. There was no online catalogue but from emails I discovered the institute held nearly all of them. These copies were accessible and I spent a lovely time at the institute's headquarters in Evans Bay, where the windows of the staff tearoom look out on crashing waves at the harbour edge.

McCann's articles were a joy. Some were annotated lists of species recorded in particular regions of India; others were narratives of his field expeditions, written in a relaxed and readable style with frequent digressions. He was obviously never far from his study subjects. A certain kind of frog would enter his home in Bombay, and in the drier conditions after the monsoon season six or seven would "hide in vases, among clothes, behind bottles and the like with the limbs drawn up under the body as close as possible [to conserve moisture]".

In a short note in 1924 McCann described the habits of a small burrowing earth snake and the local villagers' fear of the creature. A woman who saw him pick one up "was horror-stricken at the sight and gave vent to screams which promptly brought a crowd around all shouting and gesticulating". When he asked what the problem was, he was told the snake would make a knot round his hand and he would never be able to get out of it.

In September 1940, at the coastal town of Karwar, McCann found a beautiful large egg-shaped hornets' nest hanging from thin branches about a metre and a half off the ground. He attempted to secure it for the museum by plugging the entrance with cotton wool soaked in chloroform and using smoke and flames to combat the hornets outside the nest. His own protection was a raincoat tied at his wrists, sacking over his head, and a determination to stay still when the large wasps became enraged. After two days he had to admit defeat when the tinder-dry nest caught fire.

During the Second World War there are occasional references to the conflict but McCann's career and travels seem to have continued. In October 1941, during the dry season, McCann and his family went for a short holiday at the hill station of Mount Abu in Rajasthan, where the boys were to attend school. This had been the scene of McCann's first botanical tour, in May 1916, when he had travelled with Father Blatter and a Professor Hallberg, both of whom had since died. "I became reminiscent for a while," he wrote. "My dear old companions returned from the grave, and in fancy we were again together on these hills ..." His days were now filled with long walks and detailed observations of the plants, insects, mammals and birds, as well as study of the local amphibians and reptiles. He noticed a decline in some birds and an increase in mongooses, and wanted to see "the fauna of this island in the desert preserved in its proper balance".

In July of 1942 or '43 McCann returned to the Abu Hills, this time soon after the monsoon rains. There was lush green growth everywhere and things looked "altogether different to the drab grey and khaki of the dry weather". On the day he and Eleanor arrived, their son Carl came to greet them with a three-foot monitor lizard over his shoulder as a pet. "'Suffering cats! Another naturalist in the house,' said Ma. 'Haven't I enough with one?'"

McCann noted that a "grievous fault" of his own childhood had been catching snakes, and it was still one. In one article he described himself as "a man with a job and a hobby in the same profession". "I am happy," he wrote. "A naturalist is never alone in the world; he can always converse with Nature, where others, less fortunate than himself, are lost."

Opening the jars of the McCann collection was intriguing as Jason and I gradually came face to face with a selection of snakes, lizards, frogs and toads of India. McCann had labelled many specimens with the species name, or at least the genus, but many of the frogs, especially the smaller ones, had been left unnamed. Little was known then about this branch of Indian fauna. I was able to put names to some, guided by recent books and technical articles, but this project stalled when several newly published guides to Indian herpetofauna could not be viewed online and were not available for purchase or library interloan in New Zealand.

The McCann collection boasted some warty-looking Himalayan newts and a small selection of caecilians, strange legless burrowing amphibians that resemble earthworms. Other worm-like animals were the small burrowing "blind snakes", which have no use for eyes since they live underground and extend their burrows by means of their shovel-shaped snout. There were sea snakes with a flattened tail adapted for swimming, and shield-tailed snakes with a flattened shield-like tip of their tail that could be used to block a burrow. There were a few chameleons, with their bizarre eyes and grasping feet, and numerous prickly spiny agamid lizards, a family well represented in Australia. Among these were small flying lizards that have the generic name *Draco*, Latin for dragon. These creatures have short triangular heads and a flap of skin folded away along each side of their body. They live in trees and are able to launch themselves and glide remarkable distances by extending the flaps of skin to a wing-like shape supported by ribs on each side.

There were also two giant flying frogs, *Rhacophorus maximus*, a male and a female collected at Changchang Pani in the Naga Hills of Assam on February 22, 1930. The specimens, bright green in life, were now a dark bluish colour.

This lineage of frogs has extensive webbing between the long fingers and toes. They live in trees, and if they leap they spread their digits so the webbing creates air resistance and their fall is slowed to a glide. Besides northern India, giant flying frogs are found in parts of China, Tibet, Thailand and Vietnam. They live in wet forests. During breeding they lay their eggs in nests of foam placed in vegetation over a pond. When the tadpoles hatch they fall into the water below. The males are smaller: McCann's female had a snout-to-vent length of 105 millimetres and the male only seventy-three.

In 2012 I gave a short gallery talk on the McCann collection as part of a summer holiday programme at the museum. I handed the small group of adults and children some laminated photographs showing Charles McCann and a selection of the species, live, represented in his collection. On a trolley I had some of the real specimens in their jars, including the giant flying frogs. As my pièce de résistance

I removed a couple of specimens from their jar on to a tray so the children could see them closely. Jason confessed he had been sceptical that drab specimens in jars could interest the public but he reported looks of astonishment and delight on the children's faces. McCann's specimens were proving their worth eighty years on.

Seeking Pacific skinks

The jet descended and somehow out of the wide South Pacific the pilot found the speck that is the large flat island of Tongatapu in the Kingdom of Tonga. The plane skimmed low over bright blue water, a bit of rocky coastline, and then coconut palms and banana fields. I could see dusty roads and clusters of houses with thatched and iron roofs.

It was my first sight of a tropical South Seas island and I was soon hooked. Between 1986 and 1995 I made six field trips to Pacific islands. On many small islands the reptiles had not been well studied and I collected a small selection of voucher specimens for Auckland Museum's reference collection. Preserved initially in formalin and transferred to alcohol back at the museum, they made a permanent record: in the lab, with access to all the published literature, their identifications could be confirmed. Using a stereoscopic dissecting microscope I could check such features as the number of rows of scales around the middle of the body, or the number of small specialised scales—lamellae—in a row along the underside of the longest toe. By such characteristics, lizards of different species are told apart.

My first trip took me to both Tongatapu, Tonga's main island, which is densely populated and much modified, and the neighbouring island of 'Eua, which is hilly, has fewer people and retains some forest habitats. 'Eua was fifteen minutes from the main island by a Britten-Norman Islander of Friendly Islands Airways. From the air it was thrilling to preview the geography, including the forested, cliff-bound east coast. Touchdown on the grass airstrip was noisy and bumpy.

I was travelling with Anthony Wright, the Auckland Museum's botanist, who was making a collection of pressed plants. For part of the time on 'Eua we camped at a beach. We occasionally got rides in vehicles, and once we hired from a villager two tired old horses with squares of sacking for saddles. Mostly, though, we were on foot. One day we walked about sixteen kilometres to the southern coast and back. It was hot and dusty. Passing a row of palms with low coconuts, we debated how to get some. I embraced the trunk for support while Anthony stood on my shoulders and tried to knock down the coconuts with a long stick. Totally engrossed, we suddenly became aware of a Tongan man on horseback on the track behind us. Perhaps they were his trees. I stood ready to face the music. With hardly a word he dismounted, shinned up the nearest palm, and pulled two coconuts to the ground. He slithered down the tree, opened the coconuts with his machete, presented them to us and rode off.

At this time few tourists visited 'Eua and life was simple. A group of students at an agricultural college invited us to their hut to share a meal of a boiled, starchy root crop, garnished with greens that had been cooked with a little fatty meat. It was my first taste of boiled cassava root, the raw material of tapioca, also known as manioc, or manioke in Tongan, and the staple diet of millions of the poorest people in the Third World tropics. Another evening we were taken to see Tongan dancing at a village hall. We sat cross-legged on the wooden floor. Faint light came from kerosene-fuelled hurricane lamps hanging from the beams. The dancers, many of them labourers in the field by day, delighted the small attentive audience with their elegant synchronised movements to the haunting music. Some old men sitting nearby beckoned us to share kava with them. We moved on to their woven mat and were each given a half coconut shell of the famous liquid, which looks and tastes like dirty dishwater. Kava is mildly narcotising and it made my mouth and tongue feel pleasantly numb.

Prior to my visit an unusual gecko had been reported from 'Eua but I found none. A small grey-brown forest-dweller and probably most active at night, this new species was later described by two herpetologists, one in California and one based in Fiji, and named *Lepidodactylus euaensis*. Just a year later, three specimens were presented to Auckland Museum by Dieter Rinke, a German ecologist based in Tonga. Then in 1995 the museum received from the New Zealand Ministry of Agriculture and Forestry a small collection of preserved lizards, animals its staff had intercepted in cargo imported to New Zealand over many years. When I worked through them I discovered three of the unique ‘Eua geckos: they had been found in bananas in the 1960s. While I searched for lizards on ‘Eua these rare specimens had been sitting in a storeroom somewhere in Auckland.

On the tropical islands of the South Pacific the native lizards are mostly geckos or skinks. Geckos have small granular body scales, which give their skin a dull velvety appearance, and large eyes that cannot blink. They are usually nocturnal. In contrast, the body scales of skinks are flat and shiny and the small eyes usually have moveable lower eyelids that can be closed.

Skinks are often active in daylight, the sunnier the better. In the South Seas, small brown skinks with longitudinal stripes and long bluish tails are common almost everywhere. Many islands across the region also have a large black tree-climbing skink, *Emoia nigra*. While on ‘Eua I had briefly glimpsed a large brown skink moving rapidly among the exposed roots of a tree. Its identity had me puzzled.

My second South Pacific trip was to the Ha‘apai islands of Tonga in 1988, again with Anthony. Captain Cook had anchored at the main island here, Lifuka, in 1777, and the ill-fated English privateer *Port au Prince* had put in for repairs in 1806. The Tongans burned the *Port au Prince* and murdered all the crew except for a youth, Will Mariner, who lived as a Tongan for four years until he was able to escape on a passing ship. His story, as recounted to a London physician, became a bestseller in Europe.

Lifuka is a narrow island seven kilometres long. We explored it on foot and made several trips by minibus to the island of Foa over a short causeway. In the centre of Foa we found a swamp surrounded by damaged forest and there in trees I saw three of the mysterious large brown skinks. I watched them through binoculars but couldn't catch any to examine them closely. I also saw a couple of introduced European starlings that had presumably spread north from Tongatapu, a new record for the Ha'apais. Near the Tongan capital, Nuku'alofa, I saw a white-faced heron that seemed to be the first record of this species in all of Tonga. It was possibly a straggler from the population on New Caledonia.

A family in Nuku'alofa, whom we had befriended on our first Tongan visit, arranged for us to go from Lifuka to the tiny island of Lofanga, where the head of their family was a chief. One morning we rose at three-thirty. Timote Lutui, the island's headman, tapped on the window of the guesthouse as arranged. We followed him in the dark along the unsealed main road to a small settlement, where we were shown into a wooden shack. A circle of men were huddled together in the centre around a dim lantern, smoking and talking quietly. As my eyes grew accustomed to the dimness I became aware of women and children sleeping on the floor around the edges of the room. When word came that the boat was ready, we filed in the dark along a short path to the shore. A small wooden boat, partly enclosed and powered by a diesel engine, lay beached on the mud. We climbed aboard, several men pushed the boat free, and we set off on the twenty-kilometre journey across the sea.

Eventually we skirted part of Lofanga and turned into a small bay. Men on the prow directed the skipper slowly through a dynamited gap in the reef as dark heads of coral thrust towards the surface of the clear water. We landed on a sandy beach and followed Lutui a short distance to a house in a village, where we were to stay as a guest of his daughter, Pepe, and her husband, Folau. There were four rooms. The main one had woven pandanus mats on the floor but no furniture except an old cabinet for food and utensils, and a treadle sewing machine. Pepe cooked in the back garden over a small fire fuelled by the dried flower sheaths of coconut palms. We washed from an enamel basin in a thatched shelter, near a water tank that stored run-off from the house's iron roof.

Lofanga was about two kilometres long, a kilometre wide, and nowhere higher than forty-five metres above sea level. There were no vehicles or electricity. Communication with the mainland was via a public radio-telephone powered by solar panels and lead-acid batteries. The village was surrounded by a low fence to keep the pigs away from the food gardens beyond. The island was cultivated except for a thin strip of forest around the shore.

One day in an overgrown plantation I heard high-pitched chattering in the crown of a coconut palm. Eventually through binoculars I saw my first blue-crowned lories, known as henga in Tonga, an important sighting as these pretty little parrots had disappeared from many islands in the region, probably because of predation by ship rats. On rocks at the shore I caught snake-eyed skinks, *Cryptoblepharus*, which are unusual for skinks in having eyes like those of snakes or geckos that cannot close. I would later post all the Tongan snake-eyed skinks in the Auckland Museum collection on loan to George Zug at the Smithsonian Institution in Washington D.C., who was reviewing the presence of these skinks on islands across the south-west Pacific.

Our evening meals were fish caught by Folau and starchy vegetables grown in the family's allotment. We ate with our fingers, sitting cross-legged on the floor of the main room. At the end of the meal we drank a liquid that Lutui proudly called "tea". It was poured from a teapot but there the similarity with tea ended. It seemed to be a watery, sugary form of cocoa. In the thirsty climate it became something I looked forward to. When it got very dark, Lutui lit and pumped a white-spirits lantern and got his grandchildren to entertain us with their school songs.

One afternoon as I sat on the floor of the main room there was an earthquake that seemed to go on for a long time. It rocked the house sharply and made it creak. Pepe's reaction was to strike the floor with her hand and call out angrily as though to an evil spirit. I tried not to think of tsunamis and how we were on a tiny low island in the middle of an enormous ocean.

There was an idyllic beach on Lofanga. I swam there every day but was nervous of going out far since the hole in the reef that admitted boats could surely admit sharks. I asked Lutui whether swimming was safe.

"You swim all over the lagoon," he said emphatically, waving his arms.

"But how is it safe?" I persisted.

"Shark only eat bad people," he said eventually. If I swam

in safety I was as good as Lutui believed me to be. Were I injured by a shark, some bad aspect of my character would be thus exposed. Uncertain of my degree of goodness, I kept very close to the shore.

Small children usually clustered at the doorway whenever Anthony and I were working on our specimens and field notes. Many had never seen Europeans before. They were very shy at first and a direct look was enough to send them scurrying, but as their curiosity and confidence grew the children got ever closer. One day, in the corner of my eye, I saw a little hand reach out from behind to touch my bare arm.

On Sunday after breakfast we dressed for church. Both Anthony and I had had the foresight to bring white shirts. To complete our dress as honorary Tongans, the family gave us both a white tupenu, or lava-lava, to wear and a small woven waist mat ornamented with multicoloured strands of wool. Lutui led us to the largest church in the village and we sat near the front. To one side of the altar stood evidence of the short history of time-keeping on the island: the hands of a derelict grandfather clock pointed out the hour and minute at which the clock had given its last tick-tock. Perhaps the early missionaries had brought this clock to Lofanga. Its current successor was fixed to the wall, a small battery-operated clock resplendent in the slanting shafts of light. Its plastic frame was imitation gold, and on the second hand that swept around the face with electronic jerkiness rested a plastic butterfly with purple outstretched wings.

In 1989 I went to the Vava'u islands further north in the Tongan archipelago with Shelley Sayes, who had done anthropological fieldwork in Fiji. I had found a 1987 paper in which Ivan Ineich of the Muséum national d'Histoire naturelle in Paris reported his discovery that the small blue-tailed skinks on many Polynesian islands were a mix of two "cryptic" species.

The two species are superficially very similar, yet there are good features to separate them. One skink, now called *Emoia cyanura*, prefers drier, more open habitats, its tail is usually greenish or brown rather than blue, its under parts are white, and there is a paired row of scales running along the mid-line of the back, as is typical of skinks.

The other species, *Emoia impar*, prefers shady forest habitats, its tail is often bright blue, its under parts grey, and at least some of the paired scales along the mid back are fused so that the mid-line at these points has centrally placed, single fused scales, rather than the join between paired scales.

The small striped skinks now had to be examined more closely so I could decide to which of the two cryptic species they belonged, and the existing specimens in the museum had to be re-examined and re-identified. At Ineich's request, I sent him on loan for his research around a hundred specimens of these skinks from Auckland Museum's collection.

The Vava'u group is picturesque. The big main island, with its greatly indented coastline, is hemmed in to the south by a myriad of smaller islands. Sea and land are always close in every direction, and the hills are a luxuriant green. Neiafu, the town, is tucked away towards the middle of the archipelago. It was originally built to face the sea, with many of the old buildings extending down the steep hillside to the sheltered stretch of water known as Port of Refuge. This inlet was a haven for sailing ships last century and is today a mecca for ocean-going yachts.

On our first full day we found the track to the top of Mount Talau, a big hill that dominates the landscape west of Neiafu. In the forest I saw and heard my first Tongan whistlers, beautiful sparrow-sized birds unique to the Vava'u group. The male has a bright yellow body and completely black head. Another day we took a minibus as far north as it would go and then walked south-west along one of the sprawling tentacles of land that make up the main island. It was pretty country, quite open in places, with tall coconut plantations here and there, and always the blue sea in view from the dusty road.

We passed a group of workers bent over their crops in a field. They were too far off to speak to, but when they saw us they straightened up to watch us pass. We waved and they waved back. Then a teenager among them dropped his implements and ran to join us. He reached us, panting a little, and smiled from ear to ear. His name, he said, was Samisoni—that is to say, Samson. He had very little English and we knew even less Tongan, but it transpired that he knew a track to the top of a large hill called Mount Sia Ko Kafoa. The hill was marked on my map but the way to the summit was not clear. It was a pleasant climb through plantations and scrub, and I noted more Tongan whistlers. Samisoni led, in dusty shorts and a faded yellow T-shirt, and we followed.

Back down at the main dirt road I saw one of the large brown skinks on the ground and raced about trying unsuccessfully to catch it. It climbed a tree trunk and moved on to a big overhanging branch. Samisoni quickly assessed the situation and clambered up the tree. As he moved on to the big branch the lizard went further out. Samisoni began to shake the branch. The skink was now on very thin twigs that were swinging wildly. Before long it lost its grip and fell to the road below. Although I was directly underneath I had to move frantically to catch the lively animal, ending up on all fours. The mysterious lizard was now in the calico drawstring collecting bag.

Later I caught a couple more of the brown skinks. In Samoa there was a large brown tree-climbing skink, *Emoia samoensis*, and I felt these Tongan skinks were the same or a close relative. Back in Auckland I checked the measurements and patterns of scales and concluded they were *Emoia trossula*, a species from Fiji described in 1986 and also known on the Cook Islands. My specimens were the first of the species to be reported from Vava'u.

George Zug and his colleagues collected more brown skinks in the region, studied previous museum specimens, and analysed DNA samples. The evidence finally showed there were multiple species. In 2012 the Tongan brown skinks were named *Emoia mokolahi*. The specimen Samisoni had helped me catch became a paratype for the species, a secondary specimen that backs up the holotype specimen that carries the new name. A Rarotongan specimen I caught in 1995 became a paratype of another new species, *Emoia tuitarere*, which the American team described in 2011.

Our accommodation on Vava'u was a modest guest house just south of Neiafu. A slim American woman, Virginia, ran the kitchen, her greying hair tied up in a bun. She had operated a restaurant back in the US and now, finding herself in Tonga, she was staying on for a while. Her main task was to organise the evening meal in the immense dining room, where attendance was boosted by visiting yachties. Large plates of food emerged from the kitchen over an hour or two and were passed around the large communal tables. You ate your fill for a flat charge of six dollars.

One evening Virginia announced that one of the dinner dishes was "millionaire's salad" made from fresh heart of palm. Earlier in the day she had seen some men felling a coconut palm and asked them to cut out the single growing tip buried deep within the crown. The Tongans, she said, didn't use this. With a salad dressing it made a crunchy accompaniment to the meal. You can get this dish in top restaurants in New York and London but you need to be a millionaire as the heart is sent fresh from the tropics and requires the death of a palm tree. There was a peculiar satisfaction in eating this pricey salad as part of a six-dollar meal.

At one location on Vava'u I managed to catch a small skink of a kind I had not seen before. It was a climber, heavily speckled rather than striped, and had a bright lime-green belly. It seemed to be Murphy's skink, *Emoia murphyi*. In 1991 I went to Western Samoa to carry out a reptile survey and stayed two days in O le Pupu-Pu'e National Park on the south coast of the main island, 'Upolu. There I caught a Murphy's skink basking on a tree trunk and it seemed to match the Tongan specimen. This is one of the lesser known Pacific reptiles, described in 1930 from a single skink collected in Western Samoa by the Whitney South Sea Expedition of the American Museum of Natural History and named for the museum's ornithologist, Robert Cushman Murphy.

While in Apia, the capital of Samoa, I had use of a large high-ceilinged dilapidated government house down a side road near the hospital. The houses in the street had large wild lush gardens, but enormous sawn-off tree stumps marked where trees had been felled by recent cyclones. From one such stump near my bedroom window a noisy bird screeched forth intermittently day and night. It turned out to be a banded rail, known locally as ve'a. Banded rails in New Zealand are restricted to swampy sites and so shy they are rarely seen, even by birdwatchers. On Pacific Islands they are much bolder and it was amazing to see this individual stalking briskly across the lawn and climbing among the tangled weeds on the tree stump to begin its screeching.

While making a pilgrimage to the grave of Robert Louis Stevenson at Vailima, just south of Apia, I followed the well-marked jungle path up the hill and spotted a large brown Samoan skink. These skinks escape disturbance by climbing upwards and I had made many fruitless attempts to seize one before. This unlucky individual made the mistake of climbing a shrub that ended at about my height, and after a short dash across a clearing I managed to grab it.

During my last few days in Apia I walked the streets to count mynas. These birds were introduced, no doubt from nearby Pacific islands, after being brought to Fiji from India to control insect pests. The jungle myna was established at Apia before 1965 but now the common myna, the same species as is found in New Zealand, was present too, first recorded in 1988. I closely observed over 200 mynas and found that thirteen percent were common mynas.

In 1998 I was back in Apia. This time I checked the identity of just over 600 mynas around the town and found that seventy-one percent were now common mynas, the newcomer. The presence of the two species is like a natural experiment. Common mynas appear to be winning the competitive struggle between the two birds in the Apia area.

In 1993 I went to the French territory of Wallis and Futuna, which had had no published survey of reptiles. The two islands, which lie west of the Samoas and north-east of Fiji, are 230 kilometres apart and were then connected by a one-hour flight in a de Havilland Twin Otter of New Caledonia's airline, Aircalin.

In the sleepy village of Fiua on Futuna many of the gardens had low walls. As I walked quietly along the road an unusual lizard sunning on a white-painted wall caught my eye. There was nobody about so I dropped my daypack and crept towards it, hoping to catch it. Warmed by the shimmering heat, the skink darted away.

I was surprised to look up and see an elderly woman

a few feet away. She began talking excitedly in the local language, waving her arms and tucking stray wisps of hair behind her ears. I spoke French to her but she looked back blankly and called out in the direction of a banana grove. Presently a shy young woman emerged from between the plants, her dress soiled from gardening. I was able to tell her where I was from and what I was doing. She translated into Futunan for the older woman and the information was shouted up and down the street, for by now half the village was lining the road. In my experience, Polynesian villagers generally seemed to regard lizards as of no consequence, merely a plaything for children. That an adult should spend time scrabbling about catching them must have confirmed for many the unfathomable ways of Europeans.

After interest died away I was given leave to search all the nearby walls and gardens. After forty-five minutes I had caught three of the lizards, which seemed to be Murphy's skinks. I was the first to report this species from the island, although I found that two museums in Europe had unreported specimens that had been collected earlier. It turns out that the species had been named in 1899 by a German herpetologist based on two specimens from Tonga. The 1930 description of *Emoia murphyi* from Western Samoa was unnecessary. We must now call the lizard by its older name, *Emoia tongana*.

On both Wallis and Futuna I received assistance from French expatriate officers in Le service de l'économie rurale et de la pêche. On Futuna I was invited to a party one evening and met about half of the sixty or so Europeans on the island. Several tried to explain a political problem that had been developing between the three sources of power on Futuna—the customary chiefs, the Catholic church and the French administration. My French couldn't cope with the details and I understood the situation only when, during a stay in Nouméa on the journey home, I read an article about it in the local newspaper.

Two foreign priests had persuaded the two kings of Futuna, who each held sway over half the island, that a recent damaging earthquake in which five people were killed had been caused by a lack of respect for the Sabbath. The kings ruled that commercial transactions and public recreational activities such as picnics and sports matches were prohibited on Sundays. The French administration opposed this because church and state were separate in the French Republic and all its citizens, including Futunians, were free to do what they wanted on Sundays.

One king duly backed down but the other would not. Unbeknown to me, an ultimatum from the administration to lift the prohibition had expired on the fourth night of my stay. The recalcitrant monarch stayed firm so the administration disconnected power to his part of the island. I went to sleep unaware that youths were smashing every window in the town's administration building a few kilometres away and entering the home of the administrator, where they struck him.

On Sunday morning, François, an English teacher, collected me in his beaten up car to go birdwatching. It was raining heavily as we headed towards the airstrip, stopping at bays so I could scan with binoculars and count wading birds. There was a hole in the floor of the car through which

I could see the muddy road. At the airstrip we were watching kingfishers along the perimeter road when a small plane emerged from low cloud and landed. There were only a few flights a week and this one was unscheduled. In the distance we saw people emerging from their homes and running towards the airport building to find out what was happening. There had been more trouble. Five gendarmes had flown in from Wallis after youths had entered the gendarmerie, near where I was staying, and threatened the French gendarme and his local staff.

That evening I needed a walk. As I set off towards the town a Frenchman called out from a window. Hadn't I heard about the trouble? The agitators might take hostages. I should stay indoors. Later there was a knock at the door of the small apartment I was renting for the week. It was the landlord, a member of a wealthy pro-French Futunian family. I eventually understood he was worried about damage to his property and wanted me to keep the porch light on.

The next day was my last on Futuna. It was raining but I walked to town to see for myself the damage to the administration building, where only a few nights earlier I had stepped over the low gate and been able to catch house geckos attracted to the outside lights. Some bystanders told me that two men had been arrested and taken to New Caledonia. Later, back at the airport on Wallis, I saw a military transport plane standing to one side. It had brought soldiers from Nouméa but in the end they were not needed. The second king lifted the Sunday prohibitions. He and his fellow king then invited the administrator to drink kava with them and pardon the aggressors.

Rarotonga revisited

An old jar in the collection storeroom contained four birds preserved in alcohol. The lid was rusty and the fluid was the colour of dark urine. After animals are preserved in spirit, fat often leaches into the fluid, discolouring it. Working at a large stainless-steel sink, I got the lid off. When I ran the tap and poured the yellow alcohol into the sink, it immediately went milky as the dissolved fat formed an emulsion with the tap water. The birds tumbled out, along with a square of paper on which was written in pencil "T.F.C. Rarotonga".

I had just begun my career at Auckland Museum and was rehousing the collection of land vertebrates that were preserved in alcohol. Wet collections are vulnerable to drying out so this was a top priority. The specimens needed to be transferred to fresh alcohol in new jars with rust-proof plastic lids. A proportion of the specimens had never been registered. I needed to decipher any labels, assign registration numbers, and record collecting details in the numerical register and on catalogue cards.

Rarotonga is the main island of the Cook Islands group, in the tropical Pacific 3,000 kilometres north-east of New

Zealand. The "T.F.C." presumably referred to Thomas Frederic Cheeseman, the curator of Auckland Museum from 1874 until he died in 1923 in his late seventies.

I submerged the birds in a basin of tap water to rinse them, left them a while in a plastic funnel to drain, and then squeezed each one gently between paper towels to remove excess liquid. After this I placed the birds into new jars with fresh ethanol that had been diluted with tap water to seventy percent. In the crystal-clear liquid the birds looked a picture—if you like that sort of thing—with their plumage fluffed out and details such as the scales on their legs and feet clear to see.

When, a decade later, I went to the Cook Islands to survey its reptiles, my interest was renewed in the birds I had rehoused earlier. What were these drab grey medium-sized birds? They looked to be songbirds. It seemed most likely they were 'ioi, or Rarotonga starlings, *Aplonis cinerascens*. I measured the lengths of each bird's wing, tail, beak and lower leg. All measurements closely agreed with some published figures for the Rarotonga starling.

It seems Cheeseman was in Rarotonga from May to July 1899 to collect plants. In a letter of April 15, 1898 to a Mr F.J. Moss living on the island, Cheeseman stated: "The authorities of Kew Gardens are pressing me to undertake a botanical investigation of the Cook Islands ... almost nothing is known in Europe of the plants of the group. ... I should probably bring either my wife or sister with me, to attend to the drying of the plants, so as to leave my whole time free for collecting. ... I am supposing that I and my wife could obtain decent lodgings with some of the European families—for I do not know whether you have the convenience of an hotel." This was Cheeseman's only excursion beyond New Zealand. It culminated in his report "The flora of Rarotonga, the chief island of the Cook group", which was published in 1903 in *Transactions of the Linnean Society of London*.

I was excited to find that Auckland Museum's library held a diary Cheeseman kept of his Rarotonga trip. It was a pleasure to read—neatly and charmingly written. Cheeseman had made inland excursions with a lad called Josephus. After one such trip on Friday, June 23, Cheeseman wrote: "On the way back Josephus shot two specimens of the Ioi bird." The diary also referred to Cheeseman having a tank of formalin in which specimens could be preserved. It all fitted in with the jars of grey birds.

As one of Cheeseman's curatorial successors I was delighted to visit Rarotonga almost exactly a century after he had, and for similar scientific reasons. I too kept a diary.

> TUESDAY, SEPTEMBER 26, 1995 (NEW ZEALAND). A friend picked me up at noon and drove me to Auckland Airport with my pack, carry-on bag, and plastic tube containing devices for snaring lizards. During the four-hour flight we crossed the Date Line, arriving Monday after dark. How nice it was to step out into warm evening air. In the arrivals hall a man on a platform played an electrified ukulele and sang a pan-Pacific pop song. A young woman

gyrated in a bikini top and grass skirt, smiling incessantly at the queuing passengers. At the exit an official asked each passenger for the name of their accommodation and pointed to waiting hotel staff. Paul, from my motel, stepped forward and placed an ei of scented flowers around my neck. My fellow patrons and I were soon whisked by minivan along the dimly lit coast road to the lodge. It was ten p.m. I unpacked, and feeling now a little oppressed by the heat and humidity swam in the motel pool. It was lit invitingly by submerged lights.

TUESDAY, MAY 9, 1899. *Left for Rarotonga on the steamer* Ovalau, *casting off from Queen St. Wharf* [Auckland] *at quarter to nine pm. Only six passengers on board—Madame Arnaud and her daughter, two Tahitians returning to their country after a visit to New Zealand; Mr Hemus, in the employ of Messrs Donald & Edenborough; another passenger, whose name I failed to learn; and myself and wife. The night was fine but dark, and after the steamer had rounded the North Head we both returned to our cabins.*

WEDNESDAY, MAY 10. *A beautifully fine day, with hardly a ripple on the water. Numerous albatrosses and Mollymauks* [sic] *followed in the wake of our vessel, and an occasional black petrel was also seen. ...*

FRIDAY, MAY 12. *On awaking in the morning found that we were running before a strong S.W. breeze, with sails set to help us on. Towards night the wind drew more into the west, raising a beam sea, making the* Ovalau *roll rather disagreeably. …*

TUESDAY, MAY 16. *Rarotonga was sighted about 9 o'clock. At first only the high central peak was seen; then two others appeared, standing off like two little islands. By eleven o'clock the shape of the island could be distinctly made out. About two the steamer anchored off Avarua harbour, situated at the northern end of the island. High and rugged mountains were seen to occupy the whole of the central portion, with deep gorges and ravines. Many of the peaks were rocky and one remarkable column of solid rock presented a very remarkable appearance as seen from the coast a little to the south of Avarua. It must be at least 200 ft in height. Except on these steep and rocky portions the whole island, from the sea beach to the crest of the hills, was covered with luxuriant forest. …*

The anchor had no sooner dropped than we were boarded by the health officer and the custom house officers. And then some score or two of stalwart Rarotongans—tall, lithe and well-built—climbed up on the deck to offer their services for the work of discharging the cargo. We were also surrounded by quite a fleet of little outrigger canoes—each paddled by a single man. The object of this visit was evidently fishing, for as each canoe drew up under our lee, the occupant quickly had his line and hook over the side. And notwithstanding we were anchored in ten fathoms [18 metres] *of water, we could distinctly see the bottom, and even the fish swimming to and fro.*

After our luggage had been examined and passed, we landed in the little oil launch that plies between the steamer and the shore, and soon landed on the Union S.S. Co's wharf. There were crowds of gaily dressed Rarotongan women, drawn up to see the arrivals by the steamer. We were very favourably impressed by their appearance. Tall, well-built, of graceful aspect and with much natural dignity of carriage, they welcomed us without any bashfulness or hesitation, and at the same time with a politeness and suaveness seldom seen in uncivilised man. ... We had arranged to stay with Mr & Mrs Rice, whose residence turned out to be between three and four miles distant, near a native village called Arorangi. Mr Rice welcomed us immediately on our landing, and we found he had a buggy waiting to drive us out. ... The road runs parallel with the coast, seldom being more than ½ mile from it, and often much less. In many places it is lined with groves of cocoa-nuts, and here and there we passed through the cultivations of the natives—bananas, taro, kumaras, etc. ...

TUESDAY, SEPTEMBER 26 (RAROTONGA). The motel room is spacious and comfortable, but noisy as it lies close to the road and the traffic is frequent and fast. There is much wind at the moment and a constant sound of surf crashing on the nearby reef. I crossed the road to the white-sand beach and had my first view inland to the wonderfully jagged volcanic peaks of Rarotonga with their covering of lush green forest.

WEDNESDAY, MAY 17. *Early in the morning Mr Rice drove in to Avarua to fetch my heavy luggage, containing my drying paper and collecting apparatus. On his return I at once commenced to unpack. A work table was improvised in one of the sitting rooms, and I soon had my drying paper, boards and straps, books, microscope, dissecting instruments, etc etc, arranged in a very convenient way. A little closet proved to be very suitable for storing away my collecting bottles and photographic requisites. ... Mr Rice's house is a low edifice surrounded on three sides with a broad verandah. The foundations and walls are of concrete, the lime being obtained by burning the coral rocks. The windows are large and broad, and the rooms open into one another with broad openings six feet wide or more ... Standing on the front verandah, one looks directly at the central peaks of the island—steep ridges and sharp-pointed peaks, clothed with dense and luxuriant vegetation ... On each side is the virgin forest, here largely composed of the Puka* (Hernandia peltata) *and Utu* (Barringtonia speciosa), *both of them magnificent and umbrageous* [shade-giving] *trees. From the back of the house it is only a few yards to the beach—here a wide strand of coral sand white and dazzling in the sun. A little distance from the shore—perhaps quarter of a mile—is the fringing reef which everywhere encircles the island and on which a line of snow-white breakers is incessantly rolling.*

THURSDAY, SEPTEMBER 28. During daylight, buses run every hour around the coastal road in both directions.

After breakfast I caught a counter-clockwise bus. ... I alighted in the east near the Avana Valley. This runs inland for several kilometres and I set off on foot with the sea behind me. There were houses scattered about over the agricultural land, with pigs and chickens everywhere. Where the narrow road joined the ancient inland peripheral road (the Ara Metua) I got disorientated and sought help from a man working in a taro field. He had been to school in Auckland, served in the New Zealand Army, and seemed pleased to help a visiting New Zealander.

The Avana Valley began broad with wide cultivated river flats but steadily narrowed as I proceeded along the hot dusty road. I stopped to collect lizards. Large brown skinks were sunning themselves on branches and trunks and I managed to catch one by hand. I found geckos under stones and fallen logs, and I caught several of the ubiquitous little blue-tailed skinks, *Emoia cyanura* and *E. impar*, using my metre-long snaring rod. Altogether, I collected five species.

Closer to the head of the valley, forest pressed in and began to overshadow the road, which became a rough track crossing and recrossing the stream. At one of the stream crossings an American couple in shorts and T-shirts came striding towards me. They skipped across the stepping stones and the man asked if I had been to The Needle on the cross-island track. "Oh, you must go," the woman said, "it's spiritual." A few stream crossings further on, it began to rain and I turned back.

SATURDAY, MAY 20. *In the morning I walked to the native settlement of Arorangi, distant about a mile. It is quite a large village, stretching along the roadside for a considerable distance—indeed I did not reach the end of it. On either side of the road is a low stone wall, covered here and there with a species of* Peperomia. *All the houses have little patches of cultivation in front, and I was struck with the number of tropical shrubs and flowers grown for ornament. ... The residence of the chief "Ariki" of the district—at present a female—was pointed out to me. It is a long two-storied weather boarded house with verandah and balcony in front, and has a somewhat dilapidated appearance. Next to it is a large stone church—quite an imposing edifice, with huge windows. It appeared to be undergoing repairs. ...*

THURSDAY, MAY 25. *Got Matthew Kea, a bright young boy, to go with me and carry my vasculum* [plant press]. *We walked halfway to Arorangi and then struck inland until we reached the old road of Toi, along which we walked for some distance. Saw a splendid banyan tree, with its aerial roots all fused into an enormous trunk. ...*

FRIDAY, SEPTEMBER 29. After breakfast I flagged down a bus, rode north to the main town, Avarua, and took a taxi up the Avatiu Valley to the start of the cross-island walk. At first there was a road that led past houses and cultivations, then a track through semi-open country. Finally I reached the forest proper and began the ascent. In shady pockets close to the stream there were groves of the magnificent fern *Angiopteris evecta* that Cheeseman noted in a diary entry as having fronds "quite 20 ft long". I kept up a good pace to deter the mosquitoes, and the shade in the forest was welcome as I sweated up the ridges. Finally, at the base of The Needle, there were stunning views back down to Avarua and to the south coast.

A French woman had delayed to admire the view. Mountains were her element for she was a ski instructor in the Mediterranean alps. A partial ascent of The Needle was possible but there was a warning sign for the inexperienced. The track began by skirting a steep rock with a chain fixed to it as a handhold. The woman shrugged in a theatrical French way and began the climb. I contented myself with the existing views.

SUNDAY, MAY 28. *Had an early breakfast and drove to Avarua. Met Dr Craig there by appointment, and as soon as he had disposed of his hospital work we started to make the ascent of Hikurangi* [Ikurangi], *a bald peak just to the N.E. of Avarua township. We drove a couple of miles out of the town, and then turned up a side track for half a mile or so until a little hut was reached. Here we left our buggy, and after disposing of our coats and all other unnecessary impedimenta a start was made for the mountains. … After leaving the hut, the track followed the bed of a little stream for some distance. Magnificent trees of* Inocarpus edulis *were passed, but the chief vegetation was the Au* (Hibiscus tiliaceus). *Here and there were swampy patches filled with taro. …*

At last we reached the foot of the main peak, and here, at an altitude of about 800 ft, we called a halt. … Finding it impossible to keep along the ridge, we struck along the eastern face of the peak, ascending it obliquely. Great quantities of Ascarina *were noticed—also* Fitchia *and* Fagraea, *and a considerable number of trees not yet identified. The very summit of the peak was exceedingly difficult of ascent, and but for the trees growing here and there would have been still more so. Great care had to be exercised by those in front, lest they should accidentally roll stones on those below. Towards the top, quantities of* Lycopodium flagellaria *and another species were noticed, also a curious single-leaved orchid with a spike of small greenish flowers. At last the summit was reached, my aneroid* [altimeter] *making the height to be 1575 ft. …*

Our party soon rigged a flagstaff, and a large flag was soon displayed to the gaze of the inhabitants of Avarua. And then to my surprise the whole of our natives knelt down and offered up lengthy prayers as a recognition of their successful ascent. The scene was a very curious one: seven stalwart dark-skinned Polynesians, ranged in a circle on bended knees, with bare heads and serious or even devout aspect, listening attentively while their leader prayed loudly and fervently. He evoked the care of the Almighty for each of the party not forgetting to mention myself, asking that the tohunga who had come from so far to examine the plants of Rarotonga should be strengthened and supported in all his doings.

After an hour's stay on the summit we descended, taking a more direct route … It was getting dark when we reached the hut. Here a brief stay was made while a meal of bread & cheese and bananas was made. Then a drive of half an hour's duration took us to Avarua. Here a heavy shower of rain compelled us to wait until nine o'clock, after which Dr Craig drove me out to the Rices.

Cheeseman's narrative ends abruptly after an entry for July 3. Perhaps he tired of his diary as the expedition drew to an end and the exotic surroundings became commonplace. Or having encountered persistently wet weather, he may have been too occupied drying his precious plants to spare time to write. We don't know his arrangements for leaving or of his week-long voyage back to Auckland. I myself loaded my bags into a minivan, was driven to the airport, and arrived home at the end of a four-hour flight. Back at the museum I sorted my lizard specimens, checked and confirmed identifications, assigned registration numbers and filled out catalogue cards.

As for the Rarotonga starling, it is unique to that single island. Today it is uncommon and restricted to the forested interior: it seems to have largely disappeared from coastal areas as a consequence of forest clearance, predation by introduced rats, and perhaps competition from introduced (and closely related) common mynas. The specimens Cheeseman preserved in alcohol are now precious. The usual measurements and descriptions can be taken, and they can be used for study of the internal organs and for DNA sampling. The Rarotonga starling was first described in 1871 from five specimens held in Europe. The Bishop Museum in Honolulu has nine specimens collected in 1903, and there are nine in New York and at least one in Britain. There will be few, if any, in other museums.

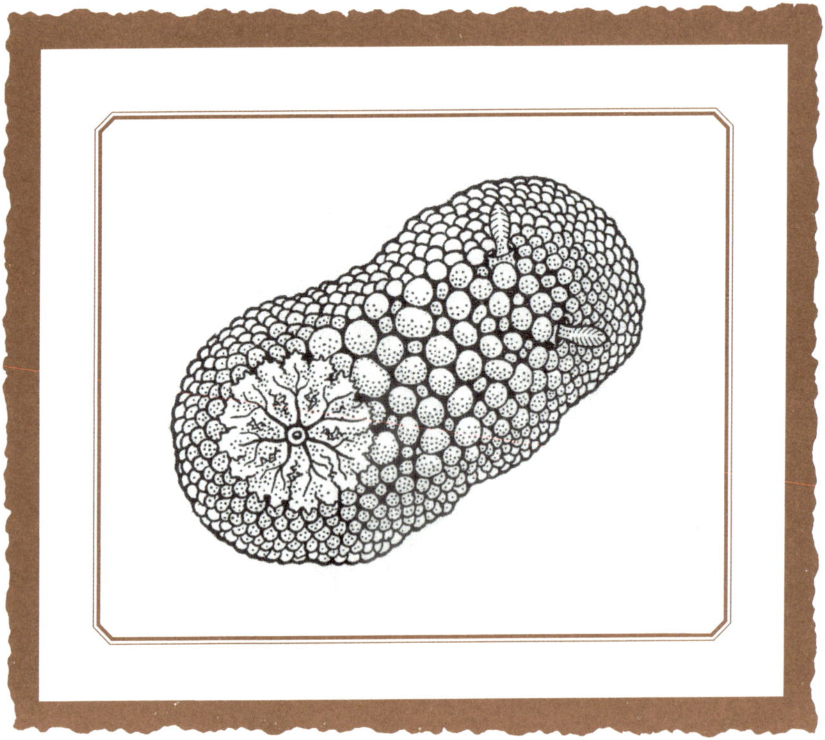

Baden Powell's sea-slug paintings

As new segments of motorway snaked across the Auckland isthmus—part of a plan to join State Highways 16 and 20—one particular section was due to slice through the edge of Mount Roskill, one of the city's volcanic cones. A walk from home took me past this hill. I was keen to know if my route beside the cone would be obliterated, so in 2005 I went to a local library to examine the environmental impact report for the project.

Skimming through the document I was pleasantly surprised to see cited, as fundamental background data for the history and importance of the hill, an article published in 1980 in Auckland Museum's *Records*. Aileen Fox, a research associate at the museum, had written the paper to record the findings of 1961 archaeological excavations at the Māori pa site on the hill.

The *Records* is Auckland Museum's regular scholarly publication. It contains academic articles, mostly on biology and anthropology. These report research results and are written by the museum's curatorial staff, or by external researchers who have used the museum collections for their studies. All the main city museums in Australia and New Zealand—and throughout the developed world—produce such volumes, which mark out the museums as being serious about scholarship. They show these institutions don't just generate new knowledge through the work of their curators and associates but also do their bit to share the burden of recording and disseminating the knowledge.

If asked where scientists and scholars publish their work, many people would think of books. But books take years to produce and are not vehicles for quick publication of the latest studies. Furthermore, most are overviews of a subject, too general to report the original data on which conclusions are based. Instead new scholarly reports, with their detailed supporting evidence, are mostly published in periodicals. These are like popular magazines: issues are frequent and each contains a group of articles that are unconnected except by the magazine's broad theme. Issue by issue, the articles build a permanent repository of information.

The invention of writing some 4,000 years ago was crucial to the development of modern civilisation. For the first time, knowledge could be accurately transmitted from place to place and across time and generations. Scholars today must publish their work if it's to have lasting value, and no study is complete until this happens. Talking about research to colleagues and making presentations to audiences will all be forgotten in time. Publication is essential for science, and when contemporary scientists are standing "on the shoulders of giants" they are standing on the published works of the giants.

Scholarly articles initially reach only a small specialist audience but they are a permanent resource, rich in information to be sought out later and reconfigured for the wider public. They can fuel much that is good and enjoyable in society, including lectures, books, films, television programmes, radio shows and exhibitions—not to mention environmental impact reports. The very stories in this book are examples of ways in which new life can be breathed into hundreds of facts first reported in a multitude of academic publications.

Auckland Museum began annual publication of *Records of the Auckland Institute and Museum* in 1930. The first issue appeared on June 30 and year by year the *Records* has continued to grow into an increasingly impressive body of knowledge. By 2010 there had been seventy issues; they contained a total of 508 articles representing the work of 173 authors and co-authors. Included in this was the description of 716 new species, subspecies, genera and higher groupings—mainly of molluscs and insects—a significant contribution by the museum to the task of documenting New Zealand's biodiversity.

To give examples of the range of papers in the *Records*, there is a 1933 article by Lucy Cranwell on Cook Islands botany, a 1949 paper by Graham Turbott recording the absence of a free-living tadpole stage in a second species of native frog, a 1958 contribution by Baden Powell on Kermadec Islands molluscs, a paper by Gilbert Archey on

Māori wood sculpture in 1967, and a 1969 paper by Janet Davidson on archaeological excavations in Tonga. Jan Allo published in 1971 on the teeth of the Māori dog, or kurī, and Gerry Barton in 1987 on a rare Māori birdman kite. A 1995 paper by Bruce Hayward and colleagues outlined a systematic survey of the fauna of the sediment of Whangape Harbour in Northland; in decades ahead this will provide invaluable baseline information for assessing the health of this harbour.

New Zealand's museum serials have a special importance in publishing lengthy works that have few outlets elsewhere. These include not only biologists' detailed descriptions of new species but archaeological reports, complete with photographs, tables, site plans and drawings of found objects, and ethnologists' accounts of cultural customs and artefacts.

The *Records* has another role: it gives the museum the chance to exchange copies for the scholarly publications of other institutions. The publication has been exchanged for hundreds of kindred titles around the world, greatly enriching the Auckland Museum library.

A pleasant surprise when I joined the museum from the university world was to discover that it boasted a magnificent research library and this was just one floor above my office. As a research project proceeded I would come across references to articles—often in obscure journals—that could prove critical in enhancing or refining my study. With the excitement of a chase, I would want those papers urgently. Many were the times I found that our library stocked a particular journal and would have a desired article thrillingly in my hand within ten minutes.

Increasingly today, electronic copies of papers can be found quickly on websites. However, paid subscriptions or single-article access fees are often required, and electronic coverage of early articles is still limited and patchy. Hard-copy publications in libraries remain relevant.

The importance of the *Records* exchanges goes beyond the value of publications received. Copies sent overseas project an international profile of research at Auckland Museum—and, in a way, research in New Zealand generally. In 2011 I was in Paris to examine long-tailed cuckoo specimens at the natural history museum. After doing my work on the bird skins I was free to walk around the park and admire the other museum buildings. When I came across the museum's lovely old library building I wanted to go inside, just to see and feel the ambience of the renowned and venerable library of one of the world's great natural history museums. I had not reckoned on there being a woman at a desk guarding the entrance. She was aloof and frosty. It seemed that you needed some sort of ticket or prior arrangement to enter. Taxing my elementary French to the limit, I managed to convince her I was a genuine researcher and that a glimpse of the library was a necessary reward for someone who had travelled so far to get there.

Inside, there was an impressive central void, bordered by open mezzanine floors. Everywhere, scholars were at work. In a prominent central position was a rack for the current issues of serial publications with their front covers enticingly displayed. Here, in a Paris library for all to see, was the current volume of *Records of the Auckland Museum*.

The editors of research serials vet the manuscripts submitted by authors and send them to experts in the field for independent comment. Authors receive the feedback, anonymously if the experts prefer, and then must modify their work to the editor's satisfaction. This process of peer review helps to maintain standards in content and academic rigour. Most papers need changing to some degree, and some are rejected outright.

In science the writing should be clear, direct, precise and succinct. Waffle, verbiage and circumlocution are the enemies. "The present author is of the opinion" should be replaced with "I believe". Technical terms and jargon have a place as stand-ins for longer and clumsier explanations, but should be an adjunct to simple prose and not a replacement for it. As a postgraduate student at Canterbury University, I found a book on science writing in the university library. It extolled as a model Michael Faraday's description of combustion in a candle flame in *The Chemical History of a Candle*, an account based on lectures he had given in 1848. Written all those years ago, this remains a masterpiece of clear interesting prose.

I've had the privilege of long involvement in the editing and production of both periodicals and books and I find it fascinating, especially the liaison with typesetters and the final preparations for printing or electronic distribution.

However, there are often great complexities. One project I took on was chairing the Checklist Committee of the Ornithological Society, which was to produce a fourth edition of the checklist of New Zealand birds. This is an official listing of the nearly 500 species and subspecies of birds in the New Zealand region, which includes Norfolk Island, New Zealand's subantarctic islands and the Ross Sea region of Antarctica. It lists the birds in taxonomic order, showing their classification into orders and families, and provides for each bird the recommended common name, the current Latin name, a list of synonyms (previously used Latin names), a synopsis of the bird's status in the country—native, introduced, extinct et cetera—and past and present distributions. Our committee of eight members was appointed in 2001 and the checklist, which once typeset ran to 500 pages, took nine years to produce.

When the manuscript was being finalised the publisher indicated it wanted all locality names rendered according to the latest recommendations of the Geographic Board, meaning some localities had to have compound English-Māori names. Wellington, for example, was allowed to stand, but Stewart Island had to become "Stewart Island / Rakiura" and Little Barrier Island "Hauturu / Little Barrier Island". I was prepared to go along with this, especially since the publisher was going to implement all the changes itself, and nobody on the committee objected.

Previous checklists had, where appropriate, given an "official" Māori name for each bird after the recommended English name in the heading. However, we now understood that many of the most familiar Māori bird names may have gained prominence simply because a missionary in a particular area happened to write down the name and it thereby entered early English–Māori dictionaries. There was, and is, much regional variation in Māori names and full research had yet to be done. The committee could not always give one preferred Māori name for a bird, as we did with English names, so after much discussion we decided to put the Māori names in a special appendix.

The next problem was Māori bird names that had become English names. During typesetting, the publisher indicated that, in line with its house style, it would be using macrons in Māori names throughout the text. A macron is a line above a vowel to indicate it is pronounced with a long sound. The publisher had already inserted them in place names, even though, as later emerged, there is not always complete agreement on where they should go. In the main text, only six bird names were affected: kākā, kākāpō, kōkako, pūkeko, takahē and tūī. I expected the committee members to accept the macrons. Instead I was surprised by the strength of the opposition. It was argued that the names, although of Māori origin, were now English names and did not warrant macrons.

One of the publishing staff we worked with was called Odessa, and a committee member who knew a bit of Russian gave me her name in Cyrillic so I could send her an email headed "Dear Одесса" to make the point that it was acceptable, for ease and convenience, to anglicise names taken into English from other languages. In emails and by telephone I argued the committee's position but the publisher stood firm. At that point I was prepared to give in, knowing the current determination of New Zealand publishers to follow trends that seem "correct", but at the eleventh hour a committee member came up with the argument that ours was the official checklist and our very task and purpose was to recommend names. It was for us to decide the names of the birds, and our names were not something a publisher should alter. The macrons were removed.

The indexes of names in the checklist became interesting resources in themselves. Many New Zealand birds have been given Latin names that repeat the country's name. These survive as the valid names for sixteen species. However, the rules of zoological nomenclature conserve original spellings. The index shows that four species use "novaezelandiae" (as in *Himantopus novaezelandiae* for the black stilt), four use "novaezealandiae" (for example, the North Island giant moa *Dinornis novaezealandiae*), and a majority, eight species, use "novaeseelandiae" (as for the morepork *Ninox novaeseelandiae*).

From the beginning Auckland Museum's *Records* included black and white photographs, although until 1969 the text was printed on matt paper so the photographs had to appear on separate glossy pages. But if you flick through the 1937 volume there is a surprise in store: a single coloured plate. This was the only time before volume 49 in 2014 that the budget allowed for colour.

The plate was printed from four-colour "process blocks" donated by a Guy C. Malcolm and appeared in an article by Arthur William Baden Powell entitled "New species of nudibranchiate Mollusca from Auckland waters". Sea slugs or nudibranchs—the word means naked gills—are soft-bodied marine molluscs with absent or reduced shells, much like terrestrial slugs, and they inhabit the bottom surfaces at all depths. On their back they often have a clump of exposed feathery gills towards the rear and a pair of sensory tentacles towards the front. Sea slugs are carnivores but they don't need to move fast: their prey are mostly stationary animals such as sponges.

There was no satisfactory way that Powell could preserve the shape and form of the delicate bodies of sea slugs so he drew them in life. The animals are often brilliantly coloured, so for the published plate he painted the six most important specimens this way. Their shapes are spectacular and their colours striking. Three are predominantly moss-green and one is orange-brown. Two are largely grey, but one of these has bright red blotches across its back and a yellow stripe along its sides. Three of the images depict new species that Powell described and named in the article; the particular specimens shown in the plate are the holotypes for these species.

The rules of zoological nomenclature require that whenever a zoologist describes and names a new animal they must identify a single specimen representative of the species—its holotype—and it is recommended that this be lodged in a public museum. These specially selected holotypes carry the new names with them. When new data or a better understanding forces a reassessment and redefinition of a species, whatever population of animals matches the holotype specimen must be called by the associated name, provided it is the oldest available valid name. If the three holotype sea slug specimens shown in Powell's plate, or any part of them, have not survived, Powell's original paintings may have to stand in for the types. When an image serves as "type" it is called an "iconotype". Many New Zealand bird specimens collected during James Cook's voyages, and subsequently named, are lost or were never preserved. These species are therefore "typified" by surviving paintings made by the expedition artists.

Arthur William Baden Powell, known as Baden, was born in Wellington in 1901 and in his youth developed a passion for conchology, the study of seashells. He trained as an artist and worked as a lithographer from 1918, which explains his talent for zoological illustration. In his spare time he was honorary conchologist at Auckland Museum. In 1929, when he was twenty-eight, the position became full-time and paid, making him one of the museum's first specialised curators. He held the job for thirty-nine years until he retired. During his career, he described nearly 1,200 new species or subspecies, with many of the descriptions appearing in the *Records*. His technical papers were usually illustrated with his exquisite drawings. In 1981 he was awarded a CBE for his contributions to marine science.

When I was a teenager I used to browse natural history books in the bookshop of the small country town where my family lived. Most were too expensive for my budget, but

I did buy one that was selling at the very attractive price of seventy-five cents and I have it still. *Native Animals of New Zealand* was written by Powell and published by Auckland Museum. It has a striking black and orange cover, and its ninety-six pages cover a selection of native species belonging to all our animal groups, from sponges to bats. There are short texts for each species but Powell's delightful inked line-and-stipple drawings are the drawcard.

The broad scope of Powell's book, the simple text, the clear and enchanting illustrations and the inexpensive price made it a hit from its first publication in 1947. The initial print run of 4,000 copies sold out in fourteen weeks and was followed by two reprints. A second edition in 1951 was reprinted five times, the last in 1975. The third edition, published in 1987, was reprinted in 1993. These first three editions sold a total of 147,000 copies, which surely makes it New Zealand's biggest-selling nature book. It used to sell itself. Bookshops all over the country would write to the museum for more stock, and one duty of the museum's receptionist was to parcel up the orders for despatch.

The book remains in print with a 1998 fourth edition as *Powell's Native Animals of New Zealand*. As an introduction to the New Zealand fauna and a means of identifying common animals, Powell's book has been a favourite for several generations of New Zealanders interested in nature. When I bought my copy as a teenager, and was learning so much from it, I little knew I would follow the author into a curatorial career at Auckland Museum, nor that it would fall to me to organise the revisions for the third and fourth editions.

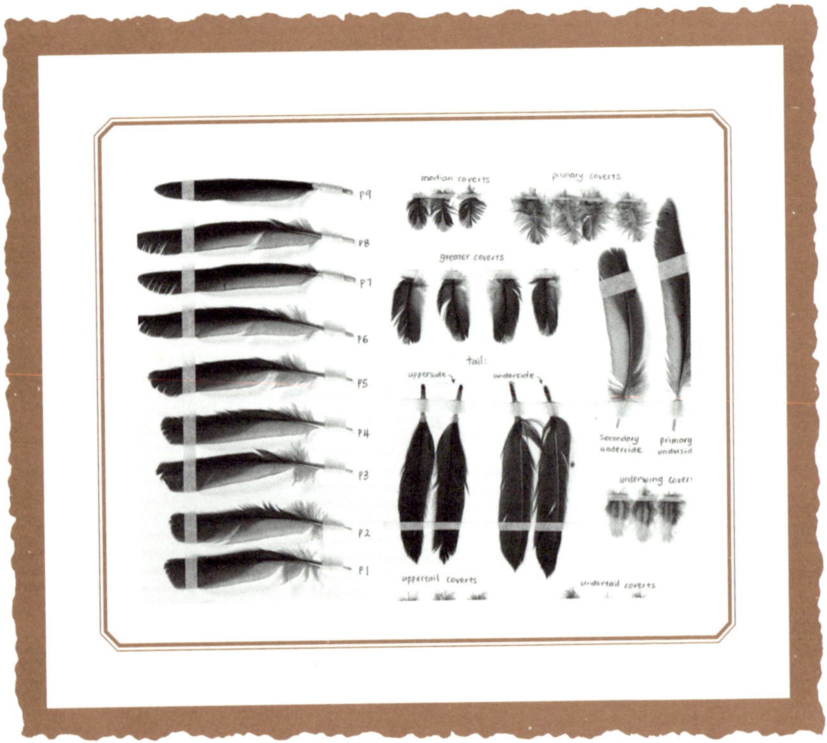
P9
P8
P7
P6
P5
P4
P3
P2
P1
median coverts
primary coverts
greater coverts
tail:
upperside
underside
secondary underside
primary
uppertail coverts
undertail coverts

Fur, feathers and frogs' legs

The third telephone call after lunch was from the Ministry of Agriculture and Fisheries. After a storm in a rural area, a member of the public had found what appeared to be a dead snake near a road washout. Leaving aside sea snakes, which occasionally straggle ashore in the north, New Zealand is one of the few countries naturally free of snakes. The ministry was concerned. If it was a snake, what kind was it and where had it come from? If it wasn't a snake, what was it?

When the specimen arrived it proved to be an Australian legless lizard. Instead of smelling of decay it smelled of methylated spirits. It was probably a preserved animal someone had thrown out, and I expect the people at the ministry were relieved to know there was no need for an expensive search for a population of exotic reptiles.

Auckland Museum has a reference collection of about 2,000 amphibians and reptiles, mostly preserved in alcohol. These cover foreign species, including snakes and legless lizards, which I was able to examine to help with the identification. Legless lizards have an external earhole;

snakes don't. Most legless lizards can close their eyes: they have a moveable eyelid. Snakes can't close their eyes: a large transparent scale is fused over each one. Many legless lizards have a small flap, or other external vestige, of the hind legs they lost relatively recently in the course of evolution; snakes have a seamless exterior surface where the hind legs would have been.

One of the many ways museums serve their local communities is by identifying mystery objects. With any luck there will be a specialist on the staff who can help. As a curator at Auckland Museum I handled hundreds of external enquiries every year and many involved giving my opinion on the likely identity of an unknown object. Many identification queries were of no lasting consequence—a birds' nest found on a back lawn, a photograph of a bird seen on holiday, or a bone picked up on a beach. However, there was a special category of enquiry, usually via official agencies, where I was asked to make identifications that could be called forensic, since potentially there were legal implications. A handful of these surfaced every year, and the first was waiting on the desk when I arrived to commence my job in 1982.

A member of the public had complained after buying a small bag of sugar and finding in it a brown feather. The manufacturer wanted to know whether the feather had entered the system in Fiji, where the sugar cane was grown and processed, or in New Zealand, where the sugar was finished and packed. Single feathers are very hard to identify unless they are distinctive in colour or shape, which this one was not. The task is made harder because museums usually don't have labelled collections of loose feathers that can be used for reference. Feathers in a stuffed bird are densely packed and overlapping and it is difficult to see the full length, shape and colour of individual ones.

From museum specimens I plucked out some feathers of likely species, but I still couldn't be sure of the identity of the brown feather from the sugar, although it was about the right colour, size and shape for a wing feather—one of the so-called "secondaries" from midway along a wing—of a common myna. This introduced Indian species is established both in Fiji and in the northern part of New Zealand, so even determining it was a myna feather would not answer the question of where it had entered the sugar. The issue was never resolved.

Pulling out feathers from study-skins to help with identification impressed on me the need for a better solution. Several years later I received a written request for feathers of New Zealand birds from a private collector in Germany. He included photocopies of sheets on which feathers had been arranged and stuck down. This gave me the idea that selected feathers from an individual bird could be removed, arranged and mounted for use as a reference. Around 1999 with the help of a technician, Maree Johnston, I made a tentative start on a collection of feather-sheets for Auckland Museum. We used the museum's herbarium materials, mounting feathers on the same large white cards and using the same archival-quality adhesive tape that were used to mount dried plants.

We built up the collection to cover all the bird species that were common locally. Being able to see entire isolated feathers quickly narrows down the possible species for identification. Feather-sheets also reveal clues to the nature of individual feathers that may not be immediately obvious. For example, in life and as a study-skin the New Zealand kingfisher, *Todiramphus sanctus*, looks bright blue-green on its upper surfaces and buff-coloured below. However, a feather-sheet for this species shows mainly grey back feathers and buff side and belly feathers, with little blue evident. A close look reveals that the blue effect is due largely to the overlapping of thousands of grey dorsal feathers, each of which has merely a touch of blue at its tip.

A similar identification problem to the feather in the sugar was a complaint that a gecko was present in a can of New Zealand beer opened in Singapore. There are many kinds of small geckos in warmer parts of the world. They are fairly similar—grey or brown with dull, velvety skin and prominent eyes—and telling them apart often hinges on a close examination of the fine structural details of their fingers and toes.

Auckland Museum maintains an excellent reference collection of geckos from both New Zealand and the Asia-Pacific region. Looking at the little gecko from the beer can through a dissecting microscope I could see that, on the upper side of the toe, the claw-bearing section at the end rose from within the expanded part of the toe, like a ballerina surrounded by a tutu. This condition is characteristic of several of the Asian "house geckos"—the geckos that commonly live in and around dwellings in the tropics—

but is not found in any New Zealand gecko, where the end of the toe appears as a simple continuation of the edge of the expanded section. I could therefore report confidently that the gecko had entered the beer in Singapore after the can was opened and not in New Zealand during the canning process. The complaint must have sprung from malice or an inaccurate observation of events. Either way, the company could defend itself against liability.

The issue of geckos allegedly turning up in New Zealand foodstuffs exported to the tropics recurred a few times—for example, with New Zealand milk powder opened in the Philippines. In each case I had the same result: a check of the morphology of the gecko's foot provided evidence that got New Zealand manufacturers off the hook. In just such a way the corporate world can benefit financially from museum collections and expertise in natural history.

Many of the forensic identifications I handled concerned food. Typically, somebody would find a strangely shaped piece of bone in a meat pie or takeaway meal and become concerned that it was evidence of something other than the expected cattle beast, sheep, pig or chicken. Supermarket chickens are immature, which is why they are so tender, and their bones are not fully formed and fused. In the museum's bone collection, chickens were represented by various adult bantams and other special breeds with bones that in size and shape were quite unlike the bones of the chickens we eat. I saved bones from a roasted chicken to make a complete reference set for the museum collection.

If the bones complained about came from a chicken meal, they always turned out to be chicken bones: they often looked unusual merely because the chicken had been chopped up at irregular angles. Similarly, an object from a cooked meat sausage that was feared to be a mouse's foot was not. It was about five millimetres long and appeared to have four pale claws, but it did not match reference examples of mice in the museum collection and the correct internal structures were lacking. It was probably a large nipple-like protruberance from a cow's or pig's tongue minced into the sausage meat. However, a possum's foot in frozen peas was exactly that, as was a mouse baked inside a loaf of bread and neatly sectioned crosswise by the factory's bread-slicer. What I received for examination were partially buttered slices. It seemed the sandwich-making had proceeded quite happily for a time but come to an abrupt halt when the horror of the situation had dawned.

In 1998, staff from an antiques auction house asked me to examine an ostrich egg with extensive scrimshaw decoration on the surface. At first glance it was a beautiful and impressive object with finely detailed illustrations inked in black on the pale cream-coloured egg surface. The weight and feel seemed right for an ostrich egg. But the situation changed under low magnification. In the land vertebrates lab I had a stereoscopic dissecting microscope that was in constant use for examining small objects on a tray. It was mounted on a large stable arm so it could also be swung out from the bench and used to examine the surface of large objects. As soon as a section of the egg came into focus I could see minute bubbles trapped in the surface that showed it to be a plastic fake. It had been produced with remarkable skill and it was still beautiful, but it was not the old scrimshaw it pretended to be.

I exposed a second hoax involving large eggs for auction. The catalogues were printed and the auction was due to take place in Auckland in December 2003. Items 84 and 85 purported to be largely intact moa eggs supposedly excavated from early Māori campsites in coastal Otago in the 1950s. Each was expected to sell for up to $7,000. I received a request from the auctioneers to examine them urgently, which I did just two weeks before the auction.

It so happened that I was currently studying moa eggs, having tracked down some thirty-six in New Zealand and foreign museums. Most of these I had examined and measured in person so if anyone knew what a moa egg looked like it was me. The eggs from the auction, in their size, shape, weight, feel and surface detail, seemed exactly like ostrich eggs. There were more than a dozen ostrich eggs in the museum's bird collection and Lots 84 and 85 were a close match. I measured the two eggs and the ratios of length divided by width were 1.22 and 1.23. From volume one of *Handbook of the Birds of the World*, I obtained the average dimensions of ostrich eggs, and the same ratio was 1.21. Ostrich eggs are "short oval" in shape—that is, tending towards round, a ratio of 1.00. Moa eggs are more elongated and the ratios of length to width for those I measured varied from 1.25 to 1.49.

The eggs were pulled from the auction. Their documentation included an old newspaper cutting reporting the original "discovery" of one of them. A photograph in the article showed a very rounded ostrich-type egg, so the hoax was perpetrated from the start and the vendor in 2003 was possibly an innocent party.

Still on the subject of eggs, there was a case where a woman was caught at Auckland International Airport attempting to smuggle into the country what appeared to be live parrot eggs. The eggs were strapped to her body and she deliberately smashed them when she was caught. I was asked to examine seven of the embryos to ascertain their identity.

The museum collection held a small selection of bird embryos preserved in alcohol but there were no parrots among them. I referred to illustrations in books held by the museum's library, and it was clear the embryos in question had several superficial features of the beak and feet consistent with their being parrots. Eggshell fragments were present and in all of them the colour of the shell was white, as is the case with parrot eggs.

I checked and approved a draft brief of evidence prepared by the Crown Solicitor. A few days later two court officials turned up at the museum's north entrance. As I walked through the building to meet them I ran into them on the grand stone staircase. In this spectacular setting they served me with a summons under the Summary Proceedings Act: I was to appear in the District Court as a witness for the prosecution. In due course, however, the offenders pleaded guilty and the case did not proceed. In around sixty forensic identifications during my career, I never did get to make a court appearance.

A man from the Ministry of Agriculture and Fisheries rang one day to say that an unusual live lizard had been dropped off anonymously at a private zoo. He assumed it to be a native lizard. If I were to examine it alive and confirm this, it could be released. Otherwise, it would have to be killed by the ministry and then referred to me for identification.

I met the woman from the zoo in the museum's main foyer. She had the lizard in a plastic ice-cream container. "It bites and it jumps," she warned. I suggested we take the container to my workroom where we could close the doors to confine the area and open it over the large deep sink. I lifted a corner of the lid, saw a beautiful scaly body, and hauled out a very lively little agamid lizard, which proceeded to bite me ineffectually. It was certainly not a New Zealand native, and may have been smuggled in from Australia or Asia. I reported this and the man from the ministry's initial response was "Oh, bugger!" The lizard was a disease threat and would have to be put down.

Another request from the ministry to identify a live reptile involved a snake that had been dumped at an animal rehabilitation centre in Whangarei after being smuggled into the country, probably on a yacht. It was now in the ministry's offices in central Auckland. When I entered the large room with the caged snake at centre stage, there were about a dozen people clustered around, including a television crew filming for the evening news. There was an awkward moment as they all stood back, seemingly expecting me to take over and deliver an illuminating verdict. I said it looked like some sort of non-poisonous constricting snake, but owned up to having no expertise on the subject. How could I have developed any working knowledge of snakes in a country that lacks them? I certainly wasn't going to handle the creature for various reasons, not least an awareness of the well-known adage about never appearing live on camera with children or animals.

The snake proved to be a boa constrictor. The media continued to be interested and dubbed it "Benny the Boa". In due course it was put down by a vet.

In 1985 authorities asked me to identify mangled remains found in the undercarriage bay of a DC-10 jet that had landed in Auckland after a flight from Los Angeles via Honolulu. The remains were those of a barn owl that must have roosted in the undercarriage and become trapped. I also received requests from the police. In one case of poaching I established that bones seized from a prepared meal—literally removed from a pot on a stove—belonged to a protected native bird the New Zealand pigeon, or kererū. In another case, helped by recourse to the museum's historic skeleton LM131, I confirmed that bones found under an old house were from domestic animals and were not human as feared.

In 2004 I was asked to identify raw meat packaged and labelled as "chicken thighs" on sale in an Asian grocery shop in South Auckland. The Institute of Environmental Science and Research, a government-owned body, was investigating on behalf of the Ministry of Agriculture and Fisheries and I went to the institute's laboratory to examine samples.

The meat was packed in meal-sized portions on polystyrene trays wrapped in clear plastic film. Superficially it looked just like packaged chicken meat, but on closer examination each "thigh" appeared to be the skinned and trimmed abdomen and legs of a large frog. Through the translucent flesh were apparent the very slender lower leg bones of frogs, which differ from the much thicker leg bones of chickens. I suggested the appropriate zoological wording for a brief of evidence that was being prepared by the ministry's investigating solicitor and signed it, but the defendants pleaded guilty at the eleventh hour and witnesses were not required in court. Total fines of around $6,000 were split between an individual and two companies, including the grocery store: it was found guilty of "recklessly selling unauthorised goods".

CM

Further reflections

Most New Zealanders have some forebears who lived in rural villages in the British Isles in the eighteenth century. When a child fell ill, someone in the village may have had access to a popular medical text, *Poor Man's Physician* by John Moncrief, published in 1712. If the child had falling-sickness, which we now call epilepsy, the book's advice was clear: "Take a little black sucking puppy (but for a girl take a bitch-whelp), choke it, open it, and take out the gall [gall-bladder] which hath not above three or four drops of pure choler [bile]; Give it all to the Child in the time of the fit, with a little tiletree flower water [from the lime tree, *Tilia*], and you shall see him cured … ."

Like human societies everywhere, we had built up a body of knowledge about the natural world but the knowledge was defective. It comprised a chaotic mix of accurate observation, trial-and-error learning, hearsay, faulty observation, pronouncements by authority figures, myths and superstitions. There was no mechanism to tell one from the other. Necessity forced all human societies to develop an armoury of remedies against illness using local plants and other materials, yet only a small proportion of remedies worked. Nobody knew which ones they were, so in ignorance and desperation you suffocated a puppy and gave its bile to an epileptic child.

Beginning roughly in the middle of the seventeenth century, the Enlightenment in Europe and North America launched a revolution in thinking and a leap in intellectual progress. Its proponents sought to replace ignorance, dogma and superstition—obstacles to progress—with reason and enquiry, which in turn gave rise to the modern scientific method. Previous knowledge of the world could now be scrutinised, item by item, to know for the first time what was sound and what was nonsense. Evidence of efficacy came to be required in medicine and the practice of smothering and eviscerating puppies for the sake of epileptic children fell by the wayside.

From science came technology and the two propelled the West to its initial pre-eminence in innovation, living standards and financial and military power. Science and technology are enabling other countries to follow suit. As if that isn't enough, the Enlightenment also contributed to the openness and democracy enjoyed in Western countries today.

Science can be defined in several ways, and the term loosely used to mean any body of knowledge. A commoner meaning is narrower: the systematic and organised knowledge arrived at by the special scientific method. Science is one of the most successful human endeavours and our best system for seeking out truth. All scientific findings are open to challenge and free to be revised and modified. Hypotheses are formulated, guesses on what the world, or a small part of it, may be like. Hypotheses cannot be proved, but they can have temporary status as an approximation of the truth so long as competing hypotheses are less well supported by the evidence. Scientists present their findings to their peers for scrutiny, and final results are published so as to be widely available for future examination and re-testing. The scientific method encourages openness, honesty and cooperation because deceit, self-deceit, falsified evidence and theft of others' work are soon exposed.

Only science has given us our modern material advances, yet increasingly it has come under public suspicion and is today under much attack. Delight in its splendour and inspiration gave ground from the middle of the twentieth century to misgivings: perhaps science dehumanised, was hijacked by dictators, caused major global problems such as nuclear accidents and chemical pollution? But science itself is of course neutral: it is its application that can be for good or evil.

I came to the museum world from university biology departments, where everyone was engaged in science and knew and understood its importance. In museum circles I began to mix professionally with people in the arts and humanities, and encountered subtle opposition to science, or viewpoints that were suspicious or mocking. An exhibition, "Enduring Nature", that I saw at Auckland Art Gallery in 2004 featured detailed cut-out shapes of insects on a wall. The label explained that the artwork was, in the curator's eyes, "a tale perhaps of the doomed human attempt to classify, to name and to control the natural world, and hence pin down its stories!"

In a glass case nearby, a volume of *The Zoology of the Voyage of H.M.S. Erebus and Terror* was open at a beautiful natural history plate. The accompanying caption read: "The conquest of New Zealand nature was attempted by ... scientists who, as Anne Salmond [in her book *The Trial of the Cannibal Dog*] explains, were 'classifying and collecting plants, animals, insects and people. ... Wild nature ... was brought under the calm controlling gaze of Enlightenment science.'"

To me, the naturalists on James Cook's eighteenth-century voyages of discovery, and on Her Majesty's Ships *Erebus* and *Terror* under James Clark Ross in their intrepid exploration of the Southern Ocean from 1839 to 1843 were heroes: in dangerous and uncomfortable conditions they took some of the earliest steps towards understanding biodiversity in our part of the world. But it seems others have a different view in which the work of natural historians is some sort of flawed and ill-advised attempt at domination and control.

These attitudes seem to have formed from a vogue among some academics for "post-modernism". Post-modernism ultimately challenges all claims to knowledge and holds that there are no facts, only interpretations. Some seem comforted by philosophies that knock science off its pedestal, among them "cultural relativism", the idea that science is simply a cultural pursuit and one world view among many that are all equally valid. Some see scientists as an élite pushing scientific methods to strengthen its own standing and wrest power from the rest of society.

In July 2015 *The New Zealand Herald* reported that in Tanzania since 2000 at least seventy-five albino adults and children have been murdered and at least sixty-two others grievously injured: their body parts are sought by traditional witch doctors for use in charms said to bring wealth and good fortune. I wonder if this is one of the world views that the cultural relativists will refuse to judge against the scientific surety that there is no known mechanism by which success in life can be enhanced by charms containing human remains? As the scientist and author Richard Dawkins put it: "An African tribe might believe that the moon is an old cooking pot thrown up into the sky, but that doesn't get you to the moon. Science does."

One contemporary threat to public understanding of biology, especially in the United States, is "scientific creationism". For religious purposes, alternative explanations for life on Earth other than evolution by natural selection are dressed up as science in an attempt to garner credibility. These interpretations, although poorly supported by evidence, are presented as truth in certain museums and textbooks.

In New Zealand there is a push to insert Māori mythical and spiritual views of the natural world into natural history. Unthinkingly, this forcing of cultural relativism on to biology undermines science by implying that scientific information on our plants and animals is optional. Where natural science forms part of a larger encyclopaedic museum, it needs to present itself on its own terms, not qualified or diluted with competing cultural or religious interpretations of nature. To do otherwise is to risk the museum's credibility.

In October 2007 representatives of natural history institutions from thirty-six countries met in Paris to discuss the global environmental crisis. It was the tercentenary of the birth of Georges-Louis Leclerc, Comte de Buffon, one of the founders of the modern science of natural history. The delegates released the Buffon Declaration, which emphasised the fourfold contribution of museums, including botanical gardens and zoos, to the sustainable management of ecosystems and therefore the survival of people. First, museums are the main repositories of the scientific samples that enable us to understand life's variety. Secondly, they contribute to research on the structure and dynamics of living things past and present. Thirdly, they help build capacity to tackle environmental challenges. And fourthly, they engage with civil society as a forum for consideration of issues crucial to our common future.

Natural history is enjoying mass popularity, at least in the West: international wildlife documentaries screening in prime time on New Zealand's commercial television channels are testament to that. A United States survey released in 2001 showed that over seventy-one million Americans, a quarter of the population, watched birds, and that birdwatching was the fastest-growing outdoor activity in North America.

Natural history collections in museums have been called libraries of life. The very purpose of the preserved specimens is to help society investigate, understand and enjoy one of the most intriguing of all scientific topics—the world's amazing biodiversity. The stories in this book are about specimens in natural history collections, and the biology, history and people connected to those specimens. I hope they have excited you about the wonder of such collections, shown that the institutions housing and caring for them are modern and relevant, and illustrated the unrivalled power of science as humanity's best hope for achieving accuracy and finding truth.

Illustrations

Registration numbers

Key specimens relevant to the stories in this book are listed below. They are held at Auckland Museum unless otherwise indicated.

Henry Ward's articulated human skeleton: LM131

Charles Adams' king penguin: LB587 (skeleton)

Tūī at the Smithsonian Institution, Washington D.C.: USNM 109192 (exchanged from Auckland Museum); 257636 (with personalised Adams label)

The unburnt moa egg: LB4016

Tokerau Beach moa eggs: LB4003 (whole); LB4005 (incomplete)

Walter Mantell's eggshell fragments: LB4013 (Taranaki); LB4014 (Ōamaru)

Bust of Richard Owen: M2579

Long-tailed cuckoos: LB8968 (egg removed from the oviduct of E39); LB8981 (E39, preserved as a spread wing)

John McLean's sample of sheep's wool: LM935

Major Buddle's Canton Island rats: LM12–13.

Major Buddle's Canton Island brown booby eggs: LB5309–10, LB13110–1. (Auckland Museum also has Buddle's eggs of two other species of boobies.)

Huia nest: Canterbury Museum, Av2743

Huia egg: Museum of New Zealand Te Papa Tongarewa, OR7640

Huia presented to Auckland Museum by Walter Buller, 1881: LB9214

Fossil takahē bones: LB7981

Fossil sea lion bone: LM759 (a well-preserved humerus that makes a good example).

Hen and Chickens Islands rats: LM9, LM31–32

Bush wren skins: LB4683–4

Charles McCann's giant flying frogs: LH3271–2

Brown skink from Sia Ko Kafoa, Tonga (paratype of *Emoia mokolahi*): LH1312

Rarotonga starlings: LB3716–9

Brown skink from Rarotonga (paratype of *Emoia tuitarere*): LH1869

Further reading

The Book of the Huia, W.J. Phillipps: Whitcombe and Tombs, Christchurch, 1963.

Canton Island, Robert Cushman Murphy, Alfred M. Bailey and Robert J. Niedrach: Denver Museum of Natural History, Denver, 1954.

"A catalogue of moa eggs (Aves: Dinornithiformes)", B.J. Gill: *Records of the Auckland Museum*, volume 43, 2006.

"Charles Francis Adams: Diary of a young American taxidermist visiting New Zealand, 1884–1887", B.J. Gill: *Archives of Natural History* 41: Edinburgh University Press, 2014.

"Deep secrets: Discovering New Zealand's tropical past", Trevor Worthy, Jenny Jones and Thomas Simpson: *New Zealand Geographic* 107, 2011. [Covers fossil birds at St Bathans, Central Otago.]

Extinct Birds of New Zealand, Alan Tennyson and Paul Martinson: Te Papa Press, Wellington, 2006.

Extinction and Biogeography of Tropical Pacific Birds, David W. Steadman: University of Chicago Press, Chicago, 2006.

"The fall of reason in the west", R. Koch and C. Smith: *New Scientist* 2557, 2006. [On the philosophy of science.]

"From treasure to tragedy—the sad tale of Blanche Halcombe", S. Hoskin: *http://pukeariki.com*

"Giants of science: Three statues with New Zealand connections", Richard Wolfe: *Art New Zealand* 152, 2014–15. [Includes Richard Owen's bronze statue.]

"History of Walter Buller's collections of New Zealand birds", J.A. Bartle and Alan Tennyson: *Tuhinga 20: Records of the Museum of New Zealand Te Papa Tongarewa*, 2009.

"How humans dared to know", A.C. Grayling: *New Scientist* 2666, 2008. [On the philosophy of science.]

Lands of Sun and Spice, Ian Turbott: Fast Books, Sydney, 1996. [Turbott's account of his postings in the Pacific.]

"The legacy of Big South Cape Island", Don Merton: *Forest and Bird* 313, 2004

The Lost World of the Moa: Prehistoric Life of New Zealand, Trevor H. Worthy and Richard N. Holdaway: Indiana University Press, Bloomington, 2002.

"Mantell, Walter Baldock Durrant: 1820–1895. Public servant, politician, naturalist", M.P.K. Sorrenson. In *The Dictionary of New Zealand Biography, Volume 1, 1769–1869*, W.H. Oliver (ed.): Allen & Unwin, Wellington, 1990.

"Natural obsessions: The Blanche Halcombe collection", Andrew Moffat. In *Flashback: Tales and Treasures of Taranaki*, Andrew Moffat (ed.): Huia Publishers, Wellington, 2012.

Richard Owen: Victorian Naturalist, Nicolaas A. Rupke: Yale University Press, New Haven, 1994.

Walter Buller: The Reluctant Conservationist, Ross Galbreath: Government Printing Office, Wellington, 1989.

Working for Wildlife: A History of the New Zealand Wildlife Service, Ross Galbreath: Bridget Williams Books in association with the Historical Branch, Department of Internal Affairs, Wellington, 1993. [Covers early transfers of rare birds to safe islands.]

Other sources

Introduction

"Historic DNA reveals contemporary population structure results from anthropogenic effects, not pre-fragmentation patterns", Lisa N. Tracy and Ian G. Jamieson: *Conservation Genetics* 12:2, 2011. [DNA study of yellowheads.]

Secrets of the shining cuckoo

"Birds collected during the Whitney South Sea Expedition.19, Notes on the bronze cuckoo *Chalcites lucidus* and its subspecies", Ernst Mayr: *American Museum Novitates* 520, 1932.

"The migration of the New Zealand bronze cuckoo, *Chalcites lucidus lucidus* (Gmelin)", H. Barraclough Fell: *Transactions of the Royal Society of New Zealand* 76, 1946–47.

The unburnt egg

"Description and conservation of a probable moa's egg (Aves: Dinornithiformes)", B.J. Gill, and S. Cooper: *Records of the Auckland Museum* 38, 2001.

"Scrambling for eggs", Bruce Ansley: *New Zealand Listener*, August 20, 1990. [Covers Canterbury Museum's broken moa egg.]

Flight of the long-tailed cuckoo

"Birds collected during the Whitney South Sea Expedition. 34, The distribution and the migration of the long-tailed cuckoo (*Urodynamis taitensis* Sparrman)", Cardine Bogert: *American Museum Novitates* 933, 1937.

"Piecing together the epic transoceanic migration of the long-tailed cuckoo (*Eudynamys taitensis*): an analysis of museum and sighting records", B.J. Gill and Mark E. Hauber: *Emu* 112, 2012.

Booby eggs and a solar eclipse

"History, present status, and future prospects of avian eggshell collections in North America", Lloyd Kiff: *The Auk* 122, 2005.

Song of the huia
"Anatomy of head and neck in the huia (*Heteralocha acutirostris*) with comparative notes on other Callaeidae", Philip John Kennedy Burton: *Bulletin of the British Museum (Natural History), Zoology* 27, 1974.

"The molecular ecology of the extinct New Zealand huia", David M. Lambert et al.: *PLoS ONE* 4(11), 2009.

"Notes on Port Nicholson and the natives in 1839", Charles Heaphy: *Transactions of the New Zealand Institute* 12, 1880.

Seals in sand dunes
"Extinction and recolonization of coastal megafauna following human arrival in New Zealand", C.J. Collins et al.: *Proceedings of the Royal Society (Series B)* 281: 20140097, 2014. [DNA study of New Zealand sea lions.]

Ship rats of Big South Cape Island
"The Big South Cape Islands rat irruption", B.D. Bell. In *The Ecology and Control of Rodents in New Zealand Nature Reserves*, Paul Richard Dingwall, I.A.E. Atkinson, C. Hay: New Zealand Department of Lands & Survey, 1978.

"A high-resolution chronology of rapid forest transitions following Polynesian arrival in New Zealand", David B. McWethy, Janet M. Wilmhurst, Cathy Whitlock, Jamie R. Wood and Matt S. McGlone: *PLoS ONE* 9(11), 2014. [Evidence for early Māori forest clearances.]

"The legacy of Big South Cape: rat irruption to rat eradication", Elizabeth A. Bell, Brian D. Bell, Don V. Merton: *New Zealand Journal of Ecology* 40, 2016.

"Prehistoric predation of the landsnail *Placostylus ambagiosus* Suter (Stylommatophora: Bulimulidae), and evidence for the timing of establishment of rats in northernmost New Zealand", F.J. Brook: *Journal of the Royal Society of New Zealand* 30, 2000.

Acknowledgements

Museums seem to attract some of the nicest people. I thank all the staff at Auckland Museum, and colleagues in wider museum circles, for their kindness and friendliness to me, at a personal level, during thirty years. Wendy Pond helped me over many years with discussions of natural history, Pacific islands and the art of writing, and I am also grateful to Ian Mason for putting me on the trail of the Ellis family in connection with J.C. McLean's egg collection. For helpful comments on drafts of specific sections I thank Fred Brook, Nigel Prickett and Heather Rogers.

I thank the Auckland Museum library staff for the library services that supported all my research. I am grateful for Auckland Museum's C.H. Worth Memorial Fund which covered the cost of my field trip to Wallis and Futuna. My recent trips to examine specimens at New Zealand and overseas museums received funding from Auckland Museum, from the bequest of the late Mrs Dora Blackie and from Dr Mark Hauber's research grants.

I thank a reader for pointing out an error in my earlier book of museum stories, *The Owl that Fell from the Sky*. On page 71 I stated that Robert Schufeldt, who published a study of the bones of a species of New Zealand parrot, drowned in the Ohio River. It was his son of the same name who drowned (in 1892). His father lived until 1934.

I am grateful to my publisher Awa Press for its efforts and high standards in producing this book: Mary Varnham worked wonders to improve my text and Emma Wolff sought out and organised the illustrations. We all thank

Auckland Museum, especially Jane Legget, Jason Froggatt and Tom Trnski, for supporting the publication of this book in various ways.

Index

Note: **bold** numbers denote illustrations.

D

E

F

Y

Z

ALSO BY BRIAN GILL

Elegantly written and told with a warm, wry humour, these fascinating stories will open your eyes to the wonders of natural history.

STEVE BRAUNIAS,
AUTHOR OF *HOW TO WATCH A BIRD*

A gem of a book—Brian Gill's engaging prose hooks you immediately and captures the absolute delight of what it is to work in a museum.

LEO JOSEPH, DIRECTOR,
AUSTRALIAN NATIONAL WILDLIFE COLLECTION

Brian Gill's The Owl that Fell from the Sky *takes the reader on a riveting journey into the rarefied world of a natural history museum... Like the collections he describes, Gill's prose is filled with wondrous and surprising detail.*

JULIA ZARANKIN,
AMERICAN BIRDING ASSOCIATION

Stuffed mammals, mysterious amphibians, eggs of extinct birds, strange-looking snakes, unusual bones—the natural history collections of museums contain thousands of such specimens. Behind many of them lie true stories of adventure, exploration, human obsession, scientific quests and strange coincidences. Brian Gill delves into the past to reveal the background of fifteen intriguing objects in New Zealand museums. In so doing, he leads us behind the scenes into the curator's world, where a phone call from the public can lead to an exciting discovery, carefully preserved specimens may help authorities detect the invasion of alien species, and the answer to a baffling mystery can come from a tiny clue.